T0314997

A Research Agenda for Digital Politics

Elgar Research Agendas outline the future of research in a given area. Leading scholars are given the space to explore their subject in provocative ways, and map out the potential directions of travel. They are relevant but also visionary.

Forward-looking and innovative, Elgar Research Agendas are an essential resource for PhD students, scholars and anybody who wants to be at the forefront of research.

Titles in the series include:

A Research Agenda for Global Crime
Edited by Tim Hall and Vincenzo Scalia

A Research Agenda for Transport Policy
Edited by John Stanley and David A. Hensher

A Research Agenda for Tourism and Development
Edited by Richard Sharpley and David Harrison

A Research Agenda for Housing
Edited by Markus Moos

A Research Agenda for Economic Anthropology
Edited by James G. Carrier

A Research Agenda for Sustainable Tourism
Edited by Stephen F. McCool and Keith Bosak

A Research Agenda for New Urbanism
Edited by Emily Talen

A Research Agenda for Creative Industries
Edited by Stuart Cunningham and Terry Flew

A Research Agenda for Military Geographies
Edited by Rachel Woodward

A Research Agenda for Sustainable Consumption Governance
Edited by Oksana Mont

A Research Agenda for Migration and Health
Edited by K. Bruce Newbold and Kathi Wilson

A Research Agenda for Climate Justice
Edited by Paul G. Harris

A Research Agenda for Federalism Studies
Edited by John Kincaid

A Research Agenda for Media Economics
Edited by Alan B. Albarran

A Research Agenda for Environmental Geopolitics
Edited by Shannon O'Lear

A Research Agenda for Studies of Corruption
Edited by Alina Mungiu-Pippidi and Paul M. Heywood

A Research Agenda for Digital Politics
Edited by William H. Dutton

A Research Agenda for Digital Politics

Edited by

WILLIAM H. DUTTON

Emeritus Professor, University of Southern California, USA, Senior Fellow, Oxford Internet Institute, Oxford Martin Fellow, Global Cyber Security Capacity Centre, University of Oxford and Visiting Professor, School of Media and Communication, University of Leeds, UK

Elgar Research Agendas

Cheltenham, UK • Northampton, MA, USA

Published by
Edward Elgar Publishing Limited
The Lypiatts
15 Lansdown Road
Cheltenham
Glos GL50 2JA
UK

Edward Elgar Publishing, Inc.
William Pratt House
9 Dewey Court
Northampton
Massachusetts 01060
USA

A catalogue record for this book
is available from the British Library

Library of Congress Control Number: 2020931684

This book is available electronically in the **Elgar**online
Social and Political Science subject collection
DOI 10.4337/9781789903096

MIX
Paper from
responsible sources
FSC   FSC® C013604

ISBN 978 1 78990 308 9 (cased)
ISBN 978 1 78990 309 6 (eBook)
Printed and bound by CPI Group (UK) Ltd, Croydon, CR0 4YY

Contents

Contributors

Nick Anstead is an Associate Professor at the Department of Media and Communications at the London School of Economics and Political Science, UK, and the Director of the MSc in Politics and Communications. A Fellow of the Royal Society of the Arts, Nick is a frequent political commentator on national and international media.

Jay G. Blumler is an Emeritus Professor of Public Communication, University of Leeds, UK, and Emeritus Professor of Journalism, University of Maryland, USA, a Fellow and past President of the International Communication Association, and holder of a life-time achievement award from the American Political Science Association.

Andrew Chadwick is Professor of Political Communication in the Department of Communication and Media at Loughborough University, UK, where he is also Director of the Online Civic Culture Centre (O3C).

Stephen Coleman is Professor of Political Communication at the University of Leeds, UK, and Research Associate at the Oxford Internet Institute, University of Oxford, UK, where he was the first professor of e-Democracy. He is a prolific scholar and speaker on issues of new media and democratic institutions and processes.

Alexi Drew is a postdoctoral researcher at King's College London, UK, working in the Centre for Science and Security Studies. She also guest lectures at Charles University in Prague, Czech Republic. Her primary research focuses on emergent technologies and the impacts on international security dynamics and their normative processes.

Elizabeth Dubois (PhD, University of Oxford) is an Assistant Professor in the Department of Communication, and member of the Centre for Law, Technology and Society at the University of Ottawa, Canada. Dr Dubois is also a Board Member for CIVIX, a Fellow at the Public Policy Forum of Canada and former member of Assembly based at MIT Media Lab and Harvard's Berkman-Klein Center, USA. Her work examines the political uses of digital media, including media manipulation, citizen engagement, and artificial intelligence.

William H. Dutton is an Emeritus Professor at the University of Southern California, USA; an Oxford Martin Fellow in the Global Cyber Security Capacity Centre, Department of Computer Science, and a Senior Fellow in the Oxford Internet Institute (OII), at the University of Oxford, UK; and a Visiting Professor at Leeds University, UK.

Laleah Fernandez is the Assistant Director of the James H. and Mary B. Quello Center in the Department of Media and Information at Michigan State University. Previously, Dr Fernandez was an Assistant Professor in the Department of Information and Computing Science at the University of Wisconsin – Green Bay, USA, having earned her PhD in Media and Information Studies, her MA in Advertising and her BA in Journalism from MSU.

Heather Ford is a researcher, lecturer and writer, thinking and writing about digital politics, charting how the Internet and digital platforms catalyse new forms of authority, power and politics, at the University of Technology Sydney, Australia.

M.I. Franklin is Professor of Global Media and Politics at Goldsmiths, University of London, UK; former Chair of the Global Internet Governance Academic Network (GigaNet) and of the Internet Rights and Principles Coalition (IRPC), and editor of *Human Rights and the Internet*, a series on openDemocracy.

Paolo Gerbaudo is a sociologist working on the transformation of social movements and political parties in the digital age, and the Director of the Centre for Digital Culture at King's College London, UK. He is the author of *Tweets and the Streets* (Pluto Press, 2012), *The Mask and the Flag* (Hurst Publishers, 2017) and *The Digital Party* (Pluto Press, 2018).

Dave Karpf is an Associate Professor in the George Washington University School of Media and Public Affairs, USA. He is the author of *The MoveOn Effect: The Unexpected Transformation of American Political Advocacy* (Oxford University Press, 2012) and *Analytic Activism: Digital Listening and the New Political Strategy* (Oxford University Press, 2016).

Leah A. Lievrouw is Professor in Information Studies at the University of California, Los Angeles (UCLA), USA.

Wan-Ying Lin (PhD) is an Associate Professor at City University of Hong Kong. Her research areas include the political uses and effects of new media, journalism studies and health communication. Her work has appeared in the *Journal of Communication, New Media and Society, Telecommunications Policy* and *Computers in Human Behavior*, among other publications.

Florian Martin-Bariteau is an Assistant Professor of Law and Technology within the Faculty of Law, Common Law Section, and the Director of the Centre for Law, Technology and Society at the University of Ottawa, Canada. As a legal scholar, his research focuses on intellectual property, blockchain, artificial intelligence, cyber-security, secrets and whistleblowers.

Declan McDowell-Naylor is a Research Associate in the School of Journalism, Media and Cultural Studies at Cardiff University, UK, where he works on the Economic and Social Research Council (ESRC) funded project, 'Beyond the MSM: Understanding the rise of alternative online political media'. He conducted his PhD research in the New Political Communication Unit at Royal Holloway, University of London, UK.

Giles Moss is Associate Professor in Media and Communication at the University of Leeds, UK.

Ben O'Loughlin is Professor of International Relations and Director of the New Political Communication Unit at Royal Holloway, University of London, UK. He is co-editor of the journal *Media, War and Conflict*. In 2019 he was Thinker in Residence on Disinformation and Democracy at the Belgian Royal Academy.

Patrícia Rossini (PhD) is a Derby Fellow at the University of Liverpool, UK. Her research focuses on online political talk, uncivil discourse, digital campaigns and deliberation. Her work has been published by *Social Media+Society*, the *International Journal of Communication*, the *Journal of Information, Technology and Politics* and the *Journal of Public Deliberation*.

Volker Schneider holds the Chair of Empirical Theory of the State at the University of Konstanz, Germany. Prof. Dr Schneider has written extensively on policy networks, technology policy and the evolution of governance structures. He is working on a theory of the modern state and public policy, combining complexity and network theories.

Lone Sorensen is Senior Lecturer at the Department of Media, Journalism and Film, University of Huddersfield, UK. She wrote her doctoral thesis on populism and political performance in established and transitional democracies. Her research addresses the topics of populism, political performance, democratic listening on social media, and mediatization in transitional democracies.

Scott Wright is Professor of Political Communication and Journalism at

Monash University, Australia. His work focuses on political talk and deliberation online, third spaces, online news comments and e-petitions.

Xinzhi Zhang (PhD) is an Assistant Professor in the Department of Journalism at Hong Kong Baptist University. His research focuses on comparative political communication and digital journalism. His work has appeared in *Computers in Human Behavior, Health Communication, Digital Journalism* and the *International Journal of Communication*, among others.

Preface

Seminal media researchers, such as Max McCombs and Donald Shaw, famously argued that the media seldom have major effects on how we think about particular issues, but do shape what we think about. They have an important 'agenda-setting' function. It is in that spirit and expectation that the collection of chapters in this book aims to play a role in shaping the agenda for research on digital politics. Whether you are a new student considering a thesis, a senior researcher considering next steps in your research, or simply an educated person with a serious interest in digital politics, this book is designed to help you think about important issues on the horizon.

This book is one of a series focused on identifying agendas for research in a wide variety of fields. My work on this book and support of the series is based on a conviction that it is critical for researchers to have a more significant role in shaping the future of research. It is not for the lack of an agenda for research, but too often researchers are driven by sources not anchored in academia.

Many sources are outside – albeit closely linked to – academia. They include the news, such as reports on the impact of new digital media, including face recognition, or contemporary issues, including deep fakes. Funding agencies, such as national and regional research councils, are influential in establishing research agendas, as are foundations and other sources of funding for research, including the digital media industries and the big technology firms. Governments and politicians set research agendas when they develop new policy initiatives or commission major policy papers concerning the media, digital technologies or political processes. Most often, these sources may well consult with academics, but are setting the research agenda from the top down.

A more bottom-up agenda develops within academia. Faculty, their departments, and centres, and associated research communities, such as around professional associations, can help academics to formulate their own agendas. And in the digital age, social media and the Internet empower academics to network with like-minded academics around the world, and to self-publish their thoughts and working papers on blogs, websites and repositories.

Nevertheless, many faculties, particularly those working in what can begin as niche areas, such as digital politics or Internet studies, do not feel that they have adequate networks of support within their own faculty or associations. In the latter half of the 1970s, when I first started to study the politics of computing in organizations and society, I not only had few colleagues working in this area, but I also had many colleagues seeking to dissuade me from pursuing this research, saying that it was 'not political science'. However, I found respected colleagues across multiple disciplines who reinforced my own sense that the political implications of information and communication technologies, such as the Internet, would be profound.

Much later, while directing the Oxford Internet Institute (OII) at the University of Oxford, I was struck by how much many doctoral students and faculty valued the opportunity to spend time at the OII, particularly in order to be among others who shared their interest and expertise in Internet studies. So, I have learned over the course of my career to appreciate the importance of academics following their own star, but also of networking with colleagues who share their interests and research agenda; one evolving from the bottom up rather than the top down.

This was the spirit in which this book was edited and produced. My aim has been to enhance the quality and impact of research on digital politics by stimulating reflection and debate among leading academics about the best topics and approaches for the future of this field. Ideally, it will be a source of support for academics involved in, or considering work on, digital politics as an important line of research.

Central to this objective has been an effort to invite chapters from a diverse range of academics. I began by identifying authors whose work has been influential in digital politics and related fields, such as the Internet and politics, political communication, and communication and technology, and the politics of information technology. While I began with this list, I invited others to balance any conservative bias anchored in past research and to avoid a risk of reinforcing dominant research streams. This process led to the identification of a range of new scholars in this area.

My invitations did not direct any author to focus on a particular area, but asked them to use this opportunity to present their ideas on what research should be undertaken by colleagues in this new field. I was not looking primarily for new research, or a literature review, but for their personal perspective on the research that should be done, based on whatever inspiration they found in their own work, and their understanding of the history and trajectory of the field. As has been the case for the field, many of the academics in this area of research have been based in the United States or the United Kingdom, so I sought to engage scholars based in other nations from across the world, and also to encourage submissions and co-authors from early- and mid-career researchers, not just senior faculty. You will see from the list of contributors that the book benefits from a global set of authors at various stages of their careers.

I hope this book helps to encourage critical discussion of the research agenda for digital politics by fostering a stronger marketplace of ideas for lone researchers as well as teams of academics working in this field. There is no expectation that this book will affect what academics think about these areas, but it might well have a heuristic role in helping academics to determine what topics and approaches to think about as they go about their own agenda-setting for their research, their field, and for the public interest in an area of vital concern across the globe.

William H. Dutton
Oxford
May 2020

Acknowledgements

I began work on this book when I was the James H. Quello Professor of Media and Information Policy at the Quello Center at Michigan State University, USA. In 2018, I returned to my home in Oxford, and research at the Global Centre for Cybersecurity Capacity Building (GCSCC) in the Department of Computer Science at the University of Oxford, where I continued as an Oxford Martin Fellow within the Oxford Martin School. I was also able to reconnect with my former department at Oxford, the Oxford Internet Institute (OII), as a Senior Fellow. I am grateful to the Quello Center, GCSCC, Oxford Martin School and the OII for their collegial support.

Since 2018, I have also enjoyed connecting with faculty at Leeds University, while a Visiting Professor. Leeds Professor Jay Blumler was valuable to my decision to take on this volume, and his advice and reviews of the outline, contributors and selected chapters has been valuable throughout the process. Professor Stephen Coleman, formerly the first Visiting Professor of e-Democracy at the University of Oxford, encouraged me to edit this volume and contributed a superb concluding chapter. In addition, Dr Yuan Zeng at Leeds provided a timely and constructive review for which I am grateful.

I am also grateful to my editor at Edward Elgar Publishing, Rachel Downie, who first approached me about this volume, and who was patient, enthusiastic and encouraging throughout the process. She and all of her many colleagues at Elgar provided timely and expert support at all stages of the book's completion. I also wish to thank Barbara Ball, who has supported the editing of this book, helping me to ensure that all the authors moved in the direction of a common style and format, while ensuring that the authors ruled when there were differences of opinion.

Finally, I am indebted to all of the authors of the chapters and thank them for providing their personal visions of an exciting agenda for this field. Their contributions have created a truly unique and stimulating book for anyone engaged with the field of digital politics. I could not have foreseen the diverse agendas envisioned by the scholars in this rapidly evolving research area, and thank them for sharing their ideas on how their fields of study can progress.

Introduction to *A Research Agenda for Digital Politics*

William H. Dutton

Many of the big questions concerning contemporary societies are about digital politics. In this book, a number of freshly minted agendas are put forward for research that could shed light on such questions as the following: Will the Internet and related digital media bring democracy to the world, or 'kill' democracy (Bartlett, 2018)? Are digital media reconfiguring access to information and people in ways that will empower citizens? Will the technologies of search and social media enable the public to be better informed about a wider range of issues, or will they be trapped in filter bubbles and echo chambers? Have digital media undermined mediating institutions, such as the quality press, in ways that are fostering nationalist and populist sentiments? Are social media and distributed intelligence leading to the 'death of expertise' (Nichols, 2017)? Will individuals be able to protect their privacy online, or will their personal data be mined by the big tech companies (Zuboff, 2019)? Will the Internet platforms centralize more power in the big tech giants, media and governments, while undermining the promise of a distributed network of individual Internet users, such as by limiting their freedom and choice (Citton, 2019)?

Answers to such questions are too often dominated by deterministic assumptions of utopian and dystopian commentators, although the pendulum of opinion has swayed heavily to the side of dystopian perspectives. Gone are the days of utopian visions of the Internet bringing democracy to the world. Quite the contrary. We are in a clearly dystopian context in which the Internet and related digital media are viewed by many as contributing to, if not causing, serious harm in politics and society (Keen, 2015; Bartlett, 2018). Even research is increasingly framed and critiqued from partisan perspectives, undermining trust in the research enterprise itself. However, you will find that the contributors to this book are neither cheerleaders for new technology, nor are they fearmongers. You will see a genuine effort by these scholars to be thought leaders in a field that is increasingly riven with deterministic thinking and partisan suspicions.

Beginning with a few of the most fundamental issues of definition, this introductory chapter identifies some key themes of research into digital politics, along with important issues facing researchers in this field, ranging from the methodological to the ethical. It notes the growing scale and sophistication of research into digital politics as one of the most critical areas of research on the social shaping and implications of technology and society (Dutton and Dubois, 2014). More researchers should be thinking about more aspects of digital politics in ways that challenge conventional wisdom and revisit early research conclusions in light of a fast-paced set of developments already taking place or visible on the horizon.

In short, academic research should be sceptical of conventional wisdom and journalistic debate lines, whether positive or negative, and seek to inform debate over these big public issues by framing questions that can be reliably and validly addressed through high-quality research. This book sets out to frame these questions.

What is digital politics?

Everything seems to be digital and everything is political, from who controls the remote control to who wins elections. But digital politics is neither an unmanageably large set of topics for research, nor too narrowly defined. Digital politics refers primarily to an increasing tendency for political actors to interact and channel their goals, strategies, activities and messages through digital platforms and media. It also enables networked individuals to respond more often and more rapidly to those actors' messages, as well as to exchange ideas with one another in narrow, wide or multiple circles. The ways and means of doing so are numerous, varied and ever evolving with innovations in digital media, communication and information technologies and their growing range of applications.

One need not be a determinist to realize that such innovations matter. They can reinforce existing structures and relationships, but in some sectors of politics and communication the consequences of this trend can be truly transformational. Hence, there is a need for continually updating agendas of research into their use and implications. While you will find each chapter of this volume providing a unique perspective on future research, you will also find some cross-cutting themes.

Themes across multiple agendas for research on digital politics

The diverse agendas offered by the contributors to this book suggest a set of cross-cutting themes behind their own unique perspectives on digital politics.

The implications of convergence

Earlier research on the political aspects of information and communication technologies was distinct from media research, and most often focused on specific technologies such as personal computers, interactive cable, videotext and satellites, joined by rising foci around the Internet, social media, mobile Internet, artificial intelligence (AI) and algorithms. Even broad terms often had a limited focus. 'Information technology' (IT) was coined in the 1950s and generally referred to computing and data-processing systems, but as computing increasingly networked people, 'information and communication technologies' (ICTs) became a better covering term. But even this broad conception did not incorporate the analogue, linear, mass media of radio, television and film. Likewise, the concept of 'media' and later 'new media', that addressed interactive cable, videotext, and more, did not necessarily cover more basic ICTs, such as the Internet and social media.

However, as technologies of ICTs and media have been converging, all moving to digital technologies, the concept of 'digital' became a broader and more useful covering term. As you will see in this book, there is still a need to focus on research anchored in non-digital technologies (see Blumler, Chapter 3 in this volume), and most research on Internet politics and digital politics is anchored in a hybrid world of analogue and digital media (Chadwick, 2006). But digital puts together technologies of the Internet and social media with more traditional media as they all move into digital forms and online across multiple devices. For instance, next-generation online television is likely to have many elements in common with the Internet and social media, all being digital media (Noam, 2019). What are the implications of this convergence to digital media?

Fragmentation amidst concentration

Technical convergence may reinforce major patterns of national and global concentration within the telecom and big tech and Internet firms: the so-called Big Nine tech giants (Crawford, 2013; Webb, 2019), as the companies build on the economies of scale inherent in digital networks. However, technical convergence and concentration are not erasing other differences across media (telecom, television, films, news, blogs, social media, and so on), ranging from their ideology to their markets, business models and regulatory regimes, which constrain convergence simply on technical grounds (Garnham, 1996). In fact, a complementary theme of a number of contributions concerns the increasing fragmentation of messages and audiences in the new digital environment.

Media researchers have worried about the fragmentation of audiences ever since the number of television channels began to multiply. Elihu Katz (1996) captured this angst in 'deliver us from segmentation'. How could mass media help to integrate the public around common texts and ideas? But segmentation and fragmentation of the public is arguably greater in the digital media context. Networked actors and individuals are producing and consuming content across multiple platforms, different social media, competing streaming services and a myriad of news services. Fragmentation makes it even more difficult to build a common understanding of events or news, for example, or to control disinformation, as different individuals use different sources (Dutton et al., 2019). However, it also makes it more difficult for any single actor to control the public's exposure to political content; for example, when they access multiple and different sources for political information. Is this a transitional period or, despite convergence and growing concentration of the media, will political actors be more fragmented, and for the better or worse?

Following the changing technological infrastructure of digital politics

Given the pace of technical change, the study of digital politics is more subject to demands to focus on the latest technical innovation than are many other fields. In many cases, it is useful to study new technologies, if only to more critically understand what they actually entail. For example, electronic voting can follow many different designs, such as remote Internet voting or replacing punch cards at voting stations with electronic machines. What is a smart city? Could there be aspects such as surveillance that are hidden by such a catchy label? What is a 'deep fake' and how can it be detected?

Nevertheless, researchers need to ensure that their agendas do not simply chase the most recent technical innovations, but pursue more fundamental questions about the political shaping and implications of digital media. In that spirit, it is possible that no technological change has been more significant than the continued worldwide diffusion of the Internet and digital media more generally, what Nanna Bonde Thylstrup (2018) calls 'mass digitization'. Since the Internet reached over half the world's population in 2019, and nearly everyone in high-income nations, the political implications of digital media have become inescapable and increasingly significant around the world. This shift in the technological infrastructure of digital politics has helped to bring more researchers into the study of digital politics, as they realize that digital politics is the new reality, but also still in its early stages of development and use in politics and society.

Challenging assumptions of the political bias of digital technologies

From the telegraph to AI, technical changes such as the move to digital are interesting as, to paraphrase Harold Lasswell (1971), they might have implications for who gets what, when, where and how. Ithiel de Sola Pool's (1983) conception of 'technologies of freedom' captured his argument that there was a 'soft technological determinism' behind the role of digital media, as they were inherently biased toward more democratic forms of access. This assumption was still prevalent in the late 1990s, when a focus was on 'digital democracy' (Dutton, 1999: 173–93). Of course, research sought to question this conventional wisdom, but this was the dominant hypothesis about the role of the new media. However, in the early decades of the twenty-first century, multiple uses of the Internet in support of autocratic regimes undermined this perspective (Howard, 2011). Yet, as I write, social movements around the world are being orchestrated through various social media in ways that seem to resurrect an earlier vision of the network of networks as a technology of freedom.

Given such uncertainties, 'digital politics' is a useful covering term because it does not embed specific assumptions about whether digital media are biased toward more democratic (decentralized) or autocratic (more hierarchical) models of political control; it is open to findings in many alternative directions. The outcome is not prejudged, even if the aim of many researchers might well be focused on strengthening democratic institutions and processes, such as by enhancing the engagement of citizens in the political process, as this entails research on counter-arguments, such that digital media might lead to 'clicktivism' that distances citizens from real politics. Over time, the thrust of empirical research has been to support the tendency of technology to be sufficiently malleable that it can be designed and used in ways that reinforce existing structures and influence; what my colleagues and I called 'reinforcement politics'

(Danziger et al., 1982). Nevertheless, the potential and actual nature of power shifts remains a persistent element of the research agenda of digital politics.

Considering the potential for new organizational and institutional forms

The expectation of power shifts is built in part on the idea of organizational change, such as what Harlan Cleveland (1985) described as the 'twilight of hierarchy'. As you will see in several contributions to this book, there is a new transformative thesis emerging, albeit not focused on democratization or autocracy. Various authors see digital politics leading to new organizational forms, such as new types of political parties (Gerbaudo, this volume), and new threats to democratic citizenship, such as around the erosion of privacy (Zuboff, 2019; Lievrouw, Chapter 16 in this volume). As noted above, digital media might enable individuals, corporations and governments to use digital media to reconfigure access to information, people and services in often unpredictable ways, placing a major premium on empirical research into the actual implications of technological change.

These new forms are associated in some treatments with the rise of digital platforms. For example, in the early years of the Internet, electronic democracy initiatives were focused on governmental and civil society innovations, such as Santa Monica, California's first electronic city hall, the Public Electronic Network (PEN). Over time, these initiatives have been more focused on platforms, for example social media platforms, such as Facebook or Twitter, and their use by individuals, political parties and advocacy groups (Karpf, 2012; Tufekci, 2017; Gerbaudo, Chapter 4 in this volume).

Challenging technologically deterministic assumptions

The importance of empirical research on the actual use and impact of digital media is also reinforced by a resurgence of technologically deterministic expectations, such as around the idea of filter bubbles (Pariser, 2011). From such perspectives, algorithms and artificial intelligence will determine what we know and whom we know. But many contributions to this book have a strong empirical and more inductivist approach, looking at qualitative and quantitative research for the actual implications of technical change in particular contexts. As noted above, the fragmentation of digital politics makes it less likely for an individual to be trapped in a filter bubble (Dutton et al., 2019). Nevertheless, deterministic perspectives such as echo chambers and filter bubbles are intuitively appealing and can provide easy answers to burning questions, such as around the greater polarization of politics, and catch on among pundits and academics alike. But deterministic perspectives need to be challenged.

The emergence of new questions and old questions in new forms

Many questions about the political implications of media, communication and information technologies are not new, such as around questions of power shifts. But in the early years of research on the political aspects of ICTs, much was based on speculation about what could be done. For example, early work on 'teledemocracy' was focused on whether interactive cable, videotext or electronic bulletin boards might be designed to support voting or polling. Was electronic voting feasible, and how would it be done? Would the Internet provide new channels for politicians to communicate with citizens? Will electronic forums and social media enable a public sphere? Will political, economic or institutional brakes constrain the development of digital government services? Will digital divides create inequities in access?

So many contemporary issues are cases of old wine in new technological bottles. For example, rising concerns over vote tampering and the security of electronic voting systems follows early warnings about the difficulties of security and authentication of actual votes. The idea of filter bubbles is focused on whether the rules and assumptions built into search algorithms might bias the outcomes of the algorithms behind search, such as limiting a user's access to more diverse sources of information. Older research on models in policy-making looked at very similar issues, such as determining the underlying assumptions and rules governing models of the impact of urban development (Dutton and Kraemer, 1985). Were they biased in ways that supported one type of development over another, such as urban infill versus urban sprawl?

Over time, however, the capability of technologies to support tasks that might have seemed like 'blue sky' thinking in earlier years has been demonstrated, such as electronic voting, or the use of AI in search, or face recognition in identification. This will not silence debate about the adequacy of the design of these technologies, but given these advances, debate has shifted more discussion to ethical issues of what should be done. You will see this in a number of contributions to this volume, such as in the discussion of rethinking citizenship in the digital age. This is a relatively new development that is certain to become more prominent in future research.

New methods and new data

Of equal importance in a number of chapters of the book are ideas about how we answer the important questions: what are the major methodological issues that are facing the field, such as around access to appropriate and reliable data on what needs to be measured versus what is easily measured? Too often in digital and computational research, the questions often seem to follow the availability of data. Instead, research questions should be guiding the design, collection and analysis of data. But even if one wished to follow the data, access to data is becoming more challenging, such as in response to concerns over privacy and data protection.

You will see in this volume a number of concerns raised about access to data for research, but also about ethical access to data. Just as tech companies have been taken to task for violating the privacy of their users, so have researchers been held accountable for the inappropriate use and sharing of personal data, fostering new academic regimes for approving research involving personal data.

The maturing of an emerging field

Finally, as a researcher in this emerging field since the mid-1970s, the contributions to this volume demonstrate a growing sophistication, and a heightened stage of maturity of this field. There is less focus on technologies, for example, and more focus on theory. You will see a wide range of theoretical perspectives from different disciplines being drawn from by the contributors to this volume. Moreover, there is less of a separation between legacy media, new media studies and digital politics. There is less deterministic thinking. The disciplinary diversity and methodological sophistication of the field is impressive. The field has moved from a handful of interdisciplinary researchers focused on technical innovations, to an increasingly large range of scholars from multiple disciplines and a diversity of theoretical perspectives who share a commitment to the study of digital politics.

The outline of this book

The book is structured in five interrelated parts. Part I introduces the field of digital politics and key issues in its future. Entitled 'Transformations and Continuities', this introductory part begins with two chapters that make a strong case for how the significance and study of digital technologies and politics has been transformed over the last decade, while the third chapter implores us not to dismiss the contributions of past decades.

Andrew Chadwick (Chapter 1) warns us that we are facing nothing less than a 'new crisis of public communication'. Dramatic events have drawn digital media from the periphery of politics into the centre of controversies over the very future of politics and democracy. According to Chadwick, this has posed a number of unprecedented challenges to scholars in this area, but also provided new opportunities for research which he outlines on the social and psychological dynamics shaping the role of social media in opinion formation. The case for new opportunities is built on by Nick Anstead (Chapter 2), who delivers an inspiring case for research priorities in this field, identifying three key challenges, including the tension between the study of the traditional and the new media environments. He described one risk tied to the excitement surrounding digital politics: scholars entering this field can unwisely neglect a rich history of research on the media and politics that has not been made redundant by new technologies. This theme is joined by Jay Blumler's (Chapter 3) contribution, which identifies key questions, concepts and theories anchored in 'legacy research'. He shows how major aspects of legacy research remain central to digital politics.

In Part II, the chapters move to one of the most central concerns around new technology and politics: the conduct of campaigns and elections. Most of the literature on politics and new media have focused on this area (Dutton and Dubois, 2014), although past work has tended to underscore a relatively limited role and impact of new technology, far from the breathtaking expectations during the early diffusion of the Internet and Web. In stark contrast, Paolo Gerbaudo (Chapter 4) begins this section with a dramatic illustration of the transformative political implications of digital platforms shaping the nature and role of political parties. His thesis and examples provide vivid illustrations of the transformative role of digital media. Declan McDowell-Naylor (Chapter 5) continues this transformational theme, focusing on the many uses of digital campaigning by political parties, along with the challenges they raise for candidates and parties in reaching voters. He identifies key substantive and methodological issues confronting research in this area, creating a range of topics for future research on election campaigns.

Laleah Fernandez (Chapter 6) also addresses the issues raised by digital media in campaigns, but also draws from her own background in advertising and communication to suggest major areas for research. These include concerns over malicious actors and their use of bots and misinformation, and the ethical issues confronting digital advertising in politics. Wan-Ying Lin and Xinzhi Zhang (Chapter 7) shift discussion from the liberal democratic contexts of North America and Europe to the People's Republic of China, where the use of digital media is more constrained by governmental restrictions on protests and other actions that challenge governmental elites. The chapter illuminates the cut and thrust of digital politics in a more restrictive context, where censorship or other sanctions can push politics into less explicitly political contexts, such as when political views gain expression through humour and everyday use of social media. This chapter is suggestive of patterns of expression that might be less visible but no less significant in other political contexts.

Part III continues the theme of transformation but shifts the focus to institutions beyond those of campaigns and elections. Volker Schneider (Chapter 8) begins this section with a theoretically and empirically rich focus on digital-era governance and its implications for the power of the state and the very nature of policy-making. He asks whether state power will recede or have a renaissance in seeking to control and regulate digitization, and what new actors will assert themselves in the policy-making process. Giles Moss and Heather Ford (Chapter 9) follow on well from the previous chapter in addressing the rise of a new institutional form – the digital platform, such as created by the big tech companies – that would not have been treated in early research on politics and the Internet. As platforms become more powerful in shaping access to information and people, should they be more accountable? How can this accountability best be exercised? Finally, Part III ends with a contribution by M.I. Franklin (Chapter 10) on the future of human rights in a digital age, threatened by the loss of privacy and the rise of digital surveillance. In light of these challenges, she proposes a refreshingly 'radical research agenda', which she sees as central to emerging technologies that are 'environmentally sustainable'.

From institutions, Part IV moves to different modes of communication, including informational, symbolic and communicative actions. In the early years of social media, debate often focused on the relative value of 'clicktivism' versus real political participation, such as street protests. Dave Karpf (Chapter 11) tells us that we should move on to other issues around advocacy and activism that move beyond debate over clicktivism, as the distinctions between online and offline behaviour become increasingly blurred. In support, he offers four new directions for research on activism and advocacy.

Of course, many profound political issues are not only over the (re)distribution of material resources, but over symbolic issues, whether they involve demonstrations of patriotism, the use of politically charged words and phrases, or other symbolic acts. Lone Sorensen (Chapter 12) focuses on some key actors and events that have used the digital media to take symbolic politics into new forms, such as in providing the very basis for a major political campaign. In doing so, she makes the case for a performance perspective on how meanings are shaped in the digital sphere, noting the constraints on bringing such a perspective into the digital politics field. Finally, Ben O'Loughlin and Alexi Drew (Chapter 13) argue that nothing speaks louder than actions to demonstrate a commitment to cyber-security. In some respects, actions on cyber-security can become a form of symbolic politics as actors seek to frame the Internet as a 'space of war on elite and public opinion'. This chapter illustrates how digital politics can meaningfully inform the study of cyber-security and other subject areas, such as public health, that might seem remote from this politics as traditionally framed.

The final part of this book (Part V) focuses on ways in which digital politics might be reshaping democratic processes: not only forms of discourse, but also the very idea of citizenship. In the aftermath of the 2016 United States presidential election, and Brexit in the United Kingdom, concerns over the civility of political discourse have risen, with many attributing this to the Internet and rising use of social media. Patrícia Rossini (Chapter 14) focuses on the incivility of political discourse, seeking insights on how to reduce the 'toxicity' of the online public sphere, without excluding the voice of those who cross some lines of civility. Scott Wright picks up aspects of this theme but focuses on what is called a 'third space' – rather than a public sphere – and asks whether this space could be Facebook. While Facebook has become a focus of criticism, particularly in the wake of the Cambridge Analytica scandal, this platform has over 2.4 billion active users, and continues to diffuse worldwide. Might Facebook become central to building a global online community, as envisioned for nearly half a century by pioneers of the Internet?

The next two chapters focus on the changing nature of citizenship – core to democratic political systems – in the digital age. Leah Lievrouw (Chapter 16) wonders if the very essence of citizenship is under threat from the degree that major Internet platforms mine the data of individuals in ways that could undermine rather than enhance the ability of individuals to access information of their choosing, and express their views in ways that are accessible to others.

Are citizens becoming 'data subjects', controlled by what platforms can collect and mine about individuals and groups? This theme is complemented and broadened by Elizabeth Dubois and Florian Martin-Bariteau (Chapter 17), who assert that no one in advanced economies can 'opt out' of the digital realm. And, therefore, not only does the very act of being a citizen need to be reconceptualized, but also leaders need to rethink traditional policies and practices for this digital age. The authors point to research that could advance this agenda.

In the last chapter of Part V, and this book, Stephen Coleman (Chapter 18) provides a powerful call for the research community to step up to the challenges of moving the study of politics into the realities of the twenty-first century. In a digital public sphere epitomized by often toxic, angry discourse, outrage, and polarization, what can be done? Coleman senses that digital politics has indeed broken traditional forms of politics, requiring us to 're-imagine' democratic processes and the democratic public in ways that individuals can feel efficacious and be heard through what is nothing less than a 'new discourse architecture', his agenda for digital politics research.

Conclusion

These 18 chapters cover much ground, but many important topics for digital politics remain beyond the scope of one book. The politics of regulating the Internet and social media, and the household politics of digital media use, such as screen time, are not in focus, even if they are undoubtedly important. Nevertheless, these chapters aim to stimulate the debates and insights that will set an agenda for research on digital politics, influencing what researchers think about in the coming years (McCombs and Shaw, 1972).

This book presents the thinking of some of the leading experts in this rapidly evolving and maturing field. It is neither comprehensive nor omniscient. You and others will shape this agenda, as you follow and participate in one of the most dynamic and controversial fields of social and political research of this century. Fifty years after the birth of the Internet, the impact of digital media is only beginning to be fully appreciated and studied as a critical means for political actors, including a public of networked individuals – digital citizens – to interact and channel their goals, strategies, activities and messages in ways that will shape the vitality of democracy, citizenship and political participation in the digital age.

References

Bartlett, J. (2018) *The People Vs Tech*. London: Ebury Publishing.

Chadwick, A. (2006) *Internet Politics*. New York: Oxford University Press.

Citton, Y. (2019) *Mediarchy*. Cambridge: Polity Press.

Cleveland, H. (1985) 'Twilight of Hierarchy', *Public Administration Review*, Jan.–Feb., pp. 185–95.

Crawford, S. (2013) *Captive Audience*. New Haven, CT: Yale University Press.

Danziger, J.N., Dutton, W.H., Kling, R. and Kraemer, K.L. (1982) *Computers and Politics*. New York: Columbia University Press.

De Sola Pool, I. (1983) *Technologies of Freedom*. Cambridge, MA: Belkap.

Dutton, W.H. (1999) *Society on the Line: Information Politics in the Digital Age*. Oxford: Oxford University Press.

Dutton, W.H. and Dubois, E. (2014) *Politics and the Internet*, Vols I–IV. London, UK and New York, USA: Routledge.

Dutton, W.H. and Kraemer, K.L. (1985) *Modeling as Negotiating: The Political Dynamics of Computer Models in the Policy Process*. Norwood, NJ: Ablex.

Dutton, W.H. and Peltu, M. (eds) (1996) *Information and Communication Technologies: Visions and Realities*. Oxford: Oxford University Press.

Dutton, W.H., Reisforf, B.C., Blank, G., Dubois, E. and Fernandez, L. (2019) 'The Internet and Access to Information about Politics'. In M. Graham and W.H. Dutton (eds), *Society and the Internet*, 2nd edition. Oxford: Oxford University Press, pp. 228–47.

Garnham, N. (1996) 'Constraints on Multimedia Convergence'. In W.H. Dutton and M. Peltu (eds) *Information and Communication Technologies: Visions and Realities*. Oxford: Oxford University Press, pp. 103–19.

Howard, P.N. (2011) *The Digital Origins of Dictatorship and Democracy*. New York: Oxford University Press.

Karpf, D. (2012) *The MoveOn Effect: The Unexpected Transformation of American Political Advocacy*. New York: Oxford University Press.

Katz, E. (1996) 'Deliver Us from Segmentation', *Annals of the American Academy of Political and Social Science*, 546 (July), pp. 22-33.

Keen, A. (2015) *The Internet Is Not The Answer*. London: Atlantic Books.

Lasswell, H.D. (1971) 'The Structure and Function of Communication in Society'. In W. Schramm and D.F. Roberts (eds), *The Process and Effects of Mass Communication*, rev. edn. Chicago, IL: University of Chicago Press, pp. 84–99.

McCombs, M.E. and Shaw, D.L. (1972) 'The Agenda-Setting Function of Mass Media', *Public Opinion Quarterly*, 36(2), pp. 176–87.

Nichols, T. (2017) *The Death of Expertise*. Oxford: Oxford University Press.

Noam, E. (2019) 'Looking Ahead at Internet Video and its Societal Impacts'. In M. Graham and W.H. Dutton (eds), *Society and the Internet*, 2nd edition. Oxford: Oxford University Press, pp. 371–88.

Pariser, R. (2011) *The Filter Bubble*. New York: Penguin Books.

Thylstrup, N.B. (2018) *The Politics of Mass Digitization*. Cambridge, MA: MIT Press.

Tufekci, Z. (2017) *Twitter and Tear Gas*. New Haven, CT: Yale University Press.

Webb, A. (2019) *The Big Nine*. New York: Hachette Book Group.

Zuboff, S. (2019) *The Age of Surveillance Capitalism*. London: Profile Books.

PART I

Transformations and Continuities

1 Four challenges for the future of digital politics research

Andrew Chadwick

Introduction

In our post-2016 moment – after the inaccurate and misleading Facebook advertising in the Brexit referendum; after the disinformation and misinformation crisis of the 2016 United States presidential campaign; after revelations of the massive scale of automated social media activity designed to manipulate public attention during key political events, some of it sponsored by Russian intelligence agencies; after the Cambridge Analytica/Facebook data breach scandal; after the live-streamed New Zealand terrorist massacre of March 2019 – the mood among digital media researchers is one of deep pessimism. So much has been lost of the optimistic visions for democratic change that once underlay scholarship in the field. Yet, even more troubling is the view that many of the pathologies of the post-2016 crisis have always been present in some form but were too often neglected. There is currently profound uncertainty about the long-term impact of all forms of digital media on civic life, but this is especially the case for social media platforms, which for many people have become the de facto Internet.

In this chapter I argue that there are four epistemological challenges facing digital politics researchers. Addressing these challenges is one way (and I hasten to add that it is only one way among many) to better equip the field for researching the post-2016 context. I also make a broader argument, which boils down to the following: in the social sciences, there are times when it is useful to shift the focus away from institutions and organizations and towards the analysis of impulses, emotions, identities and beliefs; in other words, public opinion. In the analysis of digital media and politics, that time has come.

Twenty years of research, mostly driven by normatively pro-digital media perspectives which mainly focused on whether online 'engagement' was being sufficiently embedded in political or journalistic institutions, has obscured some important questions about the properties of that engagement and the origins and consequences of digitally shaped attitudes and behaviours more generally. I want to argue that this legacy has made it more difficult for scholars to appreciate some problematic aspects of how digital media are reshaping how public opinion is formed, and how the civic culture of liberal democracies is evolving.

Four challenges for future research

The legacy of research on digital media and politics over the last two decades presents four interrelated challenges for future work in the field. I hasten to add that I have no easy solutions to these problems. In passing, I should also say that some of my own research has almost certainly contributed to them.

Analyses of digital media and politics have tended to select cases that are progressive or pro-liberal democratic

The growth of digital politics scholarship over the last two decades has mostly been fuelled by analyses of broadly progressive or pro-democratic cases (Schradie, 2019). Probably the best examples of this were the outpourings of optimistic scholarship on the 2008 Obama presidential campaign in the United States (US), and the events of 2010–11 that prematurely became known as the Arab Spring. At the level of individual interventions, probably the most significant landmark was Shirky's (2008) highly influential book *Here Comes Everybody*. But this selection bias extends in diverse ways across all kinds of cases, from political parties to social movements, to community activism, to news and journalism. Its roots are deep, complex and manifold, but an important explanation is that much scholarship on digital media and politics emerged from an instinctive critique of the elitism of the mass broadcast media system. The explosion of the Internet at the turn of this century proved so exciting because it appeared to be sweeping all of that away. It promised to usher in new forms of politics characterized by flatter hierarchies, the empowerment of the previously powerless, and a new culture of openness, tolerance and global cosmopolitanism. So many of the animating concepts in digital politics research owe their origins to this reaction against the past: the decline of journalistic gatekeepers, the horizontality of network connections, the rise of decentralized and 'leaderless' quasi-organizations, the breaking apart of monolithic government bureaucracies, the elective affinities between digital media and the new, looser, individualistic identities of progressive postmaterialism and environmental protest, to name but a few.

These disparate themes all seemed tied together for scholars who came of age intellectually with the critiques of politics, society and media that inspired social scientists during the 1990s; those whose outlook was shaped by the collapse of the Berlin Wall and the writings of Foucault, Habermas, Castells and Bourdieu, again, to name but a few. The Internet and digital media looked like the technology of a new era of freedom, in which concentrations of power in the hands of state and corporate elites would be reduced and the authentic voices of marginalized and previously under-represented groups would find untrammelled expression in the public sphere, free from the conformity and aridity of the past.

But we are facing the reality that some of the most consequential attitudes and behaviour enabled by digital media have not been particularly pro-liberal democratic or progressive. Consider, for example, inauthentic social media expression, such as the role of automated bots in social media commentary on the televised debates during the 2016 US presidential election (Kollanyi et al., 2016) and the Brexit referendum (Bastos and Mercea, 2019). Or social media expression designed to cultivate the spread of misinformation, mutual mistrust, intolerance and hatred, such as right-wing commentators' deliberate targeting of news articles about immigration (Anderson and Revers, 2018; Quandt, 2018). There is also a misleading stereotype that progressive digital mobilization has been 'bottom-up' while conservative mobilization has been orchestrated 'from above' (Schradie, 2019: 7).

As a field, we are building the conceptual and methodological tools to come to terms with these developments, but the time is ripe for research to focus attention on the intolerant and democratically dysfunctional aspects of digital media engagement. It is important to start redressing the imbalance created by the previous tendency to focus on optimistic, pro-democratic outcomes.

Research on digital media and politics has tended to employ the engagement gaze

The second challenge is a progression from the first. It is that research on digital media and politics has tended to employ the engagement gaze. By the 'engagement gaze' I mean that much research on digital media has assumed that more engagement unproblematically creates more democratic goods for the media system and the polity. The key problem is that the engagement gaze has conditioned researchers to look for evidence of engagement and, where they find it, to celebrate it as an unalloyed good. This gaze has involved under-estimating, or in some cases simply ignoring, the importance of three factors that ought to be considered when appraising any form of engagement: (1) the substantive ideological and political goals of those who become engaged; (2) the extent to which the designed-in incentive structures of any communication environment can make it more likely that some engagement will erode liberal democratic norms of authenticity, rationality and tolerance; and (3) the likelihood that new pathways for engagement will have longer-term, negative systemic consequences for the civic culture of politics.

Boulianne's remarkable meta-analysis of 320 survey-based journal articles on digital media and political participation – more than 20 years of published research – is instructive in this regard (Boulianne, 2018). Very few of those 320 journal articles have much to say about the substantive ideological and political goals of the participation they considered. Ideology appears in only 14 per cent of the 320 studies, and almost always as an explanatory variable, divorced from any specific intentions of participation. Motivations and goals mostly go unmeasured, unhelpfully obscured by the seemingly benign neutrality of the engagement construct itself. Nor did many of those 320 articles focus on how digital media affordances, such as the algorithmic sorting of information in news feeds and other relevant technological design factors, enable and constrain engagement behaviour.[1] Boulianne's analysis covered quantitative survey research. As researchers in the field will recognize, the engagement gaze is equally common in qualitative and case study-based work.

The legacy of underplaying ideological and political goals, of potentially dysfunctional technological design, and long-term consequences, has made it more difficult for scholars of digital media and politics to adapt to the post-2016 climate. And neglecting people's goals for digital engagement, while understandable for avoiding accusations of ideological bias in academia, has opened up opportunities for individuals, groups and movements who seek to produce and circulate ideas that undermine liberal democratic norms. All of this is justifiable when engagement is treated with benign neutrality, as it is under the engagement gaze.

Research on digital media and politics has mostly been driven by the rationality expectation

The third challenge is that much of the research on digital media and politics has been driven by the rationality expectation. By the 'rationality expectation' I mean the assumption that citizens are reflective, act on the best information available in the media system, and that the best resources for that action are to be found online because the Internet has comparatively few biases and distortions impacting upon the production and circulation of political knowledge. As Hedrick et al. (2018) have recently argued in a similar vein, much scholarship about digital media and politics has been informed by the assumption of an 'earnest Internet'. This, they suggest, 'generally posits that people act rationally and in good faith; care about facts, truth and authenticity; [and] pursue ends in line with their political and social values and aspirations.'

This assumption has seldom been questioned in research on digital politics, but new research is moving beyond this complacency. It is becoming clear just how widespread trolling and other behaviour that does not fit the rationality expectation is on the so-called 'ambivalent internet' (Phillips, 2016; Phillips and Milner, 2017). The more general problem is how slippery authenticity has become in digitally mediated communication. This goes beyond the long-standing argument that self-reflexivity and 'playfulness' are important parts of online culture. However, playfulness is becoming implicated in a broader culture of generalized indeterminacy. Significant majorities of the public report that, online, it is increasingly difficult to distinguish what is authentic and sincere from what is inauthentic and insincere. This is visible in the post-2016 slump in people's trust in online news and information, as evidenced across many countries in recent survey reports by Pew Research Center (2018) and the Reuters Institute (2018). Only 44 per cent of people in the 37 countries the Reuters Institute surveyed for their 2018 Digital News Report said they trust news overall. But the crisis of trust in news found via search (34 per cent) and social media (23 per cent) is far worse. The problem is compounded when one considers that only a minority of the public trusts news that they find through search engines and social media, despite the fact that these surveys also show that search engines and social media are extremely popular gateways for discovering news. So, we are faced with the paradox that most people value encountering news via media they mostly distrust.

Moving beyond the rationality expectation makes it possible to understand these and many other aspects of online politics, from the strategically deployed techniques of 'irony' and 'satire' so often used in the expression of racism, homophobia and sexism on mainstream social media platforms, and the increasingly popular alt-right sites such as 4Chan, 8Chan and Gab, to the bizarre, insider memes of the #GamerGate movement opposed to women's influence in the video game industry (and public life more generally), and the emergence of artificial intelligence (AI)-produced 'deep fake' disinformation videos. If the last of these becomes a regular feature of political events, as is already the case with fake accounts and social media bots, one possible outcome is that citizen scepticism turns to cynicism and apathy: the withdrawal that accompanies the attitude that so little political information online can be trusted because establishing the truth is so exhausting. Fears about disinformation often hinge on whether people will be deceived by falsehoods, but the lesson of the past is that people are just as likely to become uncertain about the truth and to withdraw into the private sphere. This was an important strand of the critique of the neo-Stalinist states in Eastern Europe (e.g., Havel, 1985).

We are already seeing early signs of this culture of indeterminacy. For example, Toff and Nielsen's qualitative research in the north of England has shown that 'I don't know what to believe' has become an important response to the uncertainties of encountering news on social media and private messaging apps (Toff and Nielsen, 2018). Meanwhile, Petersen et al. (2018) have gathered survey data that show what they term a 'need for chaos' is an important motivation among those in Western democracies who share false rumours and conspiracy theories. As the authors put it, '[T]he sharing of hostile political rumors is not motivated by a desire to aid actors within the system. Instead, it is motivated by a desire to tear down the system.' Such motivations are present among about 40 per cent of the US population (Petersen et al., 2018). All of this points to an unsettling vision of the future, but to see it and address it requires relaxing the rationality expectation and coming to grips with some of the wilder frontiers of online mis- and disinformation.

Research on digital media and politics has often underestimated the trade-offs between affective solidarity and rational deliberation

The final challenge is that previous research on digital media and politics has tended to underestimate the inevitable trade-offs between affective solidarity and rational deliberation. A new wave of research is foregrounding emotion as a key force in media, politics and journalism (e.g., Papacharissi, 2015; Wahl-Jorgensen, 2018). But there is a tension in this work. As Papacharissi has convincingly demonstrated, affect interacts with the affordances of social media, particularly circulation, repetition and recursion, and plays a role in opinion formation and collective action by contributing to the social solidarity and identity that are essential precursors to political mobilization (Papacharissi, 2015). But the social force of affect online is also such that the identities from which it springs, and which it shapes and reinforces, seem to be highly resistant to challenge and subversion. Identity based on affective ties seems to be particularly difficult to dislodge online. This point underlies much of the anxiety about 'filter bubbles' and 'echo chambers', it informs much of the empirical research on misinformation and misperceptions, and as Kreiss argues, it ultimately implies the fracturing of civic epistemology, 'the basis upon which people understand and agree upon political facts and truths' (Kreiss, 2017; see also Waisbord, 2018).

Arlie Russell Hochschild's recent book *Strangers in their Own Land* (Hochschild, 2016), about conservative identity in the American South, is instructive in this regard. While Hochschild has little to say about the role of media, she reveals much about how complex, multilayered emotional substructures condition daily life and attitudes to political and economic authority. This provides a useful perspective for exploring the roots of intolerance and misinformation online.

Animating affective divides is what Hochschild terms 'deep stories'. A deep story is essentially an overarching, metaphorical sensibility. It serves as an emotionally charged basis for everyday orientations towards social, cultural, economic and political reality. The deep story feeds identity, social division and resentment toward the other side, however that side is defined. As Hochschild puts it:

> A deep story is a feels-as-if story – it's the story feelings tell, in the language of symbols. It removes judgment. It removes fact. It tells us how things feel . . . And I don't believe we understand anyone's politics, right or left, without it. For we all have a deep story.

Concerns about online echo chambers may be exaggerated, not least because the empirical evidence for these phenomena has always been mixed. But the ways in which social media platforms have positioned behavioural metrics cues and algorithmic sorting of content at the centre of their business models for garnering attention has had effects on group attitudes and behaviour. The US misinformation and bot crises provide evidence of how Facebook's news feed and Twitter's hashtags have introduced surprising new vulnerabilities, but the problem is a larger one: content that reinforces one's identity is more accessible than ever to those in society who are motivated to have their identity reinforced, even if that identity is based on democratically dysfunctional norms, such as the refusal – fuelled by misogyny, xenophobia or racism – to hear the other side. Decades of political psychology research on selective exposure have demonstrated that many people are predisposed to having their existing attitudes reinforced by their media consumption habits, but we are only in the early stages of learning about how selective exposure informs people's online production and sharing habits.

This is an example of when an understandable cognitive bias – the need to have one's views reinforced to reduce risk and uncertainty – converges with social media affordances: visible metrics cues and algorithmic curation of feeds and hashtags. This heady mix is highly conducive to building collective action through affective solidarity; and quickly, even if the information is false. But it is not so beneficial for rational deliberation and building consensus through recognition and respect for difference. The inevitable trade-offs for liberal democratic political culture caused by this tension ought to be examined more carefully in future research.

Technological affordances have played only a minor role in traditional public opinion scholarship, which has been mostly about the message, not the medium. And yet, public opinion research has paid attention to problematic aspects of how public opinion is formed, for example through the study of elite cues (e.g., Edelman, 1988; Zaller, 1992). It has also considered the limits to genuinely informed public opinion when media exposure is conditioned by partisanship, polarization and motivated reasoning. Attention to how the affordances of digital media interact with the constraints that we know shape citizens' reasoning about public affairs can update these approaches. And, of course, we should also bear in mind that social media affordances are not always what they appear to be on the surface. They are vulnerable to being exploited, often in hidden ways, by actors of various kinds who seek to distort the economy of attention and influence public opinion through subterfuge, spreading false rumours, or splicing together information from a range of different sources, some reputable, some not.

As social media increased in popularity over the last decade, attention to elite cues was mostly sidelined in favour of conceptual frameworks that focused on what seemed to be so 'new' about 'new media': individual agency, the erosion of traditional gatekeepers, and user-generated content (e.g., Bruns, 2008). But we are learning that techniques of elite persuasion, while they have been reconfigured, are still highly important for the formation of public opinion and political behaviour. An important task, then, is to identify the conditions under which democratically dysfunctional information spreads online, exposing potentially large numbers to content which – and this is crucial – many then choose to share in their own social media networks.

Conclusion

In this chapter I have argued that research on digital politics ought to be recalibrated to explain how digital media can shape public opinion in ways that are dysfunctional for liberal democratic societies. As I remarked at the beginning, there are no easy solutions, but it is also the case that this task is already under way across the social sciences. The overarching challenge we face is that disinformation, misinformation, hatred and intolerance are radically networked like never before, and the raw materials for individuals and organizations to behave in democratically dysfunctional ways are diverse and multiple.

In the context of this new crisis of public communication, a key task ahead is explaining how social, psychological and technological variables converge in ways that shape how individuals form identities and opinions about the political world. Can we bring these objects of analysis together – social and psychological variables, and the affordances of digital media – to help understand how liberal democracy is evolving as the great expansion of digital media platforms nears completion? Can we develop better understandings of how social media interact with, and potentially reconfigure, the different constraints on rational opinion formation at the individual level? Is it possible to identify the blend of cognitive biases, social identities and affordances that align to produce democratically dysfunctional behaviour that threatens the future of liberal democracy? And, finally – perhaps the biggest challenge of all – can we, as scholars, effectively intervene in public and policy debates to minimize the impact of these forces in the interests of promoting liberal democratic norms?

Acknowledgements

I thank Samantha Bradshaw, Nick Couldry, David Karpf, Sarah Anne Ganter, Philip N. Howard, Daniel Kreiss, Shannon C. McGregor, Lisa-Maria Neudert, Rasmus Kleis Nielsen, Stuart Soroka, Talia Stroud, Peter Van Aelst and Gadi Wolfsfeld for their comments on an earlier presentation of these ideas. Any errors or shortcomings in this chapter are mine.

Note

1. I am grateful to Dr Shelley Boulianne for sending me the full variable list for her 2018 article.

References

Anderson, C.W. and Revers, M. (2018) 'From Counter-Power to Counter-Pepe: The Vagaries of Participatory Epistemology in a Digital Age', *Communication and Media*, 6(4), pp. 24–35.

Bastos, M. and Mercea, D. (2019) 'The Brexit Botnet and User-Generated Hyperpartisan News', *Social Science Computer Review*, 37(1), pp. 38–54.

Boulianne, S. (2018) 'Twenty Years of Digital Media Effects on Civic and Political Participation', *Communication Research*. doi: 10.1177/0093650218808186.

Bruns, A. (2008) *Blogs, Wikipedia, Second Life, and Beyond: From Production to Produsage*. New York: Peter Lang.

Edelman, M. (1988) *Constructing the Political Spectacle*. Chicago, IL: University of Chicago Press.

Havel, V. (1985) *The Power of the Powerless: Citizens Against the State in Central Eastern Europe*. London: Routledge.

Hedrick, A., Karpf, D. and Kreiss, D. (2018) Book Review: 'The Earnest Internet vs. the Ambivalent Internet', *International Journal of Communication*, 12(8), pp. 1057–64.

Hochschild, A.R. (2016) *Strangers in their Own Land: Anger and Mourning on the American Right*. New York: New Press.

Kollanyi, B., Howard, P.N. and Woolley, S.C. (2016) 'Bots and Automation over Twitter During the First US Presidential Debate (Comprop Data Memo 2016.1)'. Computational Propaganda Project, Oxford Internet Institute. Available at http://blogs.oii.ox.ac.uk/politicalbots/wp-content/uploads/sites/89/2016/10/Data-Memo-First-Presidential-Debate.pdf/.

Kreiss, D. (2017) 'The Fragmenting of the Civil Sphere: How Partisan Identity Shapes the Moral Evaluation of Candidates and Epistemology', *American Journal of Cultural Sociology*, 5(3), pp. 443–59.

Papacharissi, Z. (2015) *Affective Publics: Sentiment, Technology, and Politics*. New York: Oxford University Press.

Petersen, M.B., Osmundsen, M. and Arceneaux, K. (2018) 'A "Need for Chaos" and the Sharing of Hostile Political Rumours in Advanced Democracies'. Paper presented to the American Political Science Association Annual Meeting, 2018. Available at https://psyarxiv.com/6m4ts.

Pew Research Center (2018) *News Use Across Social Media Platforms 2018*. Washington, DC: Pew Research Center.

Phillips, W. (2016) *This Is Why We Can't Have Nice Things: Mapping the Relationship Between Online Trolling and Mainstream Culture*. Cambridge, MA: MIT Press.

Phillips, W. and Milner, R.M. (2017) *The Ambivalent Internet: Mischief, Oddity, and Antagonism Online*. Cambridge: Polity Press.

Quandt, T. (2018) 'Dark Participation', *Media and Communication*, 6(4), pp. 36–48.

Reuters Institute for the Study of Journalism (2018) *Reuters Institute Digital News Report 2018*. Oxford: Reuters Institute. http://media.digitalnewsreport.org/wp-content/uploads/2018/06/digital-news-report-2018.pdf.

Schradie, J. (2019) *The Revolution That Wasn't: How Digital Activism Favors Conservatives*. Cambridge, MA: Harvard University Press.

Shirky, C. (2008) *Here Comes Everybody: The Power of Organizing Without Organizations*. London: Allen Lane.

Toff, B. and Nielsen, R.K. (2018) '"I Just Google It": Folk Theories of Distributed Discovery', *Journal of Communication*, 68(3), pp. 636–57.

Wahl-Jorgensen, K. (2018) *Emotions, Media and Politics*. Cambridge: Polity.
Waisbord, S. (2018) 'Truth is What Happens to News: On Journalism, Fake News, and Post-Truth', *Journalism Studies*, 19(13), pp. 1866–78.
Zaller, J. (1992) *The Nature and Origins of Mass Opinion*. New York: Cambridge University Press.

2 The future of political communication research

Nick Anstead

Introduction

In order to map out a possible agenda for future political communication research, and in particular highlight some risks and challenges facing the field, this chapter draws on two core arguments. First, that political communication research has historically been shaped by three driving factors. Specifically, these are: the evolution of communication technologies, and particularly new technologies that have become important to political communication practices; the perceived health of democratic institutions, and the quality and quantity of citizen engagement; and the research possibilities offered by methodological innovation. These drivers shape the possibilities of our field, and frame the questions that are prominent within it. As a necessary corollary, they also create limitations and shortcomings in our research. As such, many of the contemporary challenges facing the study of political communication would have been recognizable to earlier generations of researchers.

However, this chapter will also argue that the stakes for the field have become considerably higher than was previously the case. This is the second core argument of the chapter: namely, that debates about political communication are increasingly central to broader debates about politics. This is true both within the academy, and more widely among politicians, journalists and the general public. As a result of this growing centrality, political communication scholars have a hugely important and growing responsibility to shape, inform and, where necessary, temper those debates. As such, thoughtfully responding to the risks and challenges facing the field takes on a much greater significance.

The three drivers of political communication research: key questions and risks

While any field of academic study will likely have a multifaceted and hugely complex lineage, a strong case can be made that the trajectory of political communication research has been heavily shaped by three distinct factors, which can be summarized as technology, democracy and methods.

Historically, interest in questions of political communication have often gone hand-in-hand with technological innovation. Arguably, the original manifestation of this is the proliferation of televisions into citizens' living rooms, first in the post-war United States (US) and then elsewhere.[1] This was notable in discussions about the 1960 (US) presidential election debates, decades of research on the effects of television advertising and, more recently, interest in 24-hour partisan cable news channels.

Contemporary research has increasingly shifted to studying the Internet. The focus of this work has changed as the Internet has evolved, with early work looking at campaign webpages, and then moving onto email communication, and subsequently studying social media and data-driven targeting of voters. Debates triggered by technological changes have produced some of the most innovative, important and challenging work in the political communication field. However, a technological focus has also created problems. The risk of developing an overly technologically determinist view of political communication has been widely discussed (for an excellent commentary on this, see Garrett et al., 2012). Particularly, it leads to overly simplistic explanations for social phenomena that neglect the inevitable complexity and interaction of a variety of causes behind any given event.

Related to this is a risk of research becoming overly platform-specific. This can lead to a tendency to overly focus on whatever social media platform is most closely tied to the events leading the news agenda at any given moment. Such an approach runs two risks. First, it neglects the extent to which the broader media ecosystem is constructed from the interactions between a variety of newer media platforms, not to mention more traditional mass media (Chadwick, 2013). Second, it ignores the social, institutional and political dimensions that are intrinsic to the phenomena that we are seeking to explain (Wolfsfeld et al., 2013).

Discussions of what has been termed 'fake news' offer a good example of this problem. A huge amount of energy has been deployed studying the role played by social media platforms, particularly Facebook, in the creation and circulation of fake news. This is clearly a relevant part of any understanding of the phenomenon, but there are other dimensions to this problem that need to be understood to build a more holistic account of what is happening. A non-exhaustive list might also include the geopolitical aspects (Russian interference in elections in other countries); the relationship between legacy media and fake news (on this issue, see Chadwick et al., 2018); the institutional dimension, which has left electoral regulators completely unable to cope with the challenges thrown up by the problem; and the attitudes to politics held by elements of the electorate that has made them prone to believing the myths and fake news stories that are being circulated (for example, a lack of trust in politicians and political institutions).

A technological focus in political communication research also runs the risk of diverting our energies to platforms that have taken on significance in particular countries or narratives around specific events. For example, recent events in the US and United Kingdom (UK) placed Facebook at the top of many researchers' agendas. This is understandable, particularly after the Cambridge Analytica scandal revealed questions about the misuse of customer data, and practices being employed to target voters with particular political messages (Cadwalladr and Graham-Harrison, 2018b). However, in other countries, such as India, closed messaging platforms such as WhatsApp are arguably far more important (Murgia et al., 2019).

Second, the perceived health of democratic institutions has provided the essential context for debates about political communication. The contemporary manifestation of this are discussions about populism and would-be authoritarian leaders. However, similar patterns are equally true of the post-Vietnam and Watergate studies of the 1970s and 1980s, notably Michael Robinson's (1976) 'Videomalaise' theory, which argued that the way in which television presented politics increased the public's level of cynicism, or discussions about decreasing political engagement and voter turnout in the 1990s and early 2000s (e.g., Putnam, 2001).

Discussion about the health of democratic institutions and levels of engagement is one area where political communication research can make a meaningful and important contribution to contemporary debates. There are, however, risks in centring research wholly around democratic life, and particularly the problems faced by liberal democracies. One concern is disproportionate levels of panic around the robustness of democratic institutions. Political theorist David Runciman (2017), for example, has detailed a series of democratic crises in the twentieth century, arguing that the great strength of liberal democracy (in contrast to more authoritarian systems of government) is flexibility in response to such situations. By following and bolstering a narrative of a general crisis in democratic institutions, political communication scholars may inadvertently be fuelling a sense of democratic fatalism by diagnosing a crisis (or crises) so large that they seem beyond the realm of individual, institutional or societal action. However, if Runciman is right, and democratic institutions are likely to remain robust, the challenge shifts to offering ideas about how those institutions can be improved.

A second risk is that the current focus of our research is too heavily driven by the experiences of particular democratic polities. This leads to a narrowing of the questions being asked by political communication scholars. The twin traumas of Donald Trump's election to the White House and the UK's referendum vote to leave the European Union, both in 2016, have driven interest in reactionary, right-wing populist politics. While it has produced much excellent work, there are far fewer studies examining contradictory examples. Why, for example, has the German Green Party developed into such a formidable force in recent years?

It should be acknowledged that an overly narrow geographical focus is a long-standing weakness of political communication research and the discipline of communications more generally. Evidence repeatedly suggests that research published in top journals is dominated by scholars working in a few regions of the world, particularly based at institutions in the US and, to a lesser extent, some Western European countries, notably the UK and the Netherlands. Huge inequalities do exist between the Global North and Global South, although some countries in the Global South, such as India and South Africa, are generating increasing amounts of scholarship (however, the quantities are still a long way behind the established academic powers in the field). In contrast, Eastern European and Latin American scholarship remains relatively isolated from the rest of the field (for a detailed discussion of data in this area, see Waisbord, 2019: Chapter 4).

Broadening the geographic scope of the study of political communication creates new and interesting opportunities, especially in the contemporary political context. The study of populism, for example, suggests comparisons that would have been considered unusual in the past. We could learn a great deal in a comparison of the political communication practices of, for example, the Modi government in India, the Orbán regime in Hungary and the Trump White House in the US. Theoretical insights might also have great value flowing in what are, by historical standards, unusual directions. For a long time, political communication research has been debating the

'Americanization' of political communication in other countries, that is, the extent to which communication practices developed in the United States have permeated other countries. The converse may be just as important. If we want to understand contemporary US politics, we could do a lot worse than examine a hypothesis focusing on 'Latin Americanization', that is, whether the politics of the US is coming to increasingly resemble the weakly institutionalized and highly leader-focused model of politics recognizable from many parts of Latin America for much of the twentieth century.

Above all, the field requires a stronger engagement with questions of both democratic theory and practice. While the macro discourse about democratic institutions and engagement has become overly pessimistic in many countries, there are also exciting possibilities created by new and successful modes of political engagement. For example, a huge amount has been written about the UK's 2016 European Union membership referendum and subsequent political fallout. However, an alternative model of practice is provided by the 2018 Irish referendum on amending the constitution to legalize abortion. This was preceded by a Citizens' Assembly on the topic, which allowed its members to engage with experts on the issue and suggest proposed changes to the constitution, which were then submitted to the Dáil for discussion before being put to the people in a plebiscitary vote. Political communication scholars should be as interested in these positive examples, which effectively merge different models of democratic participation.

Third, methodological innovation has played a huge role in shaping political communication research. The deployment of survey methods, particularly in the US in the post-war period was hugely significant, with the American National Election Study first being deployed in 1948. Other countries developed similar research instruments in the following decades. Subsequent innovations in survey research, particularly the development of telephone and online panel surveys, have made surveys both more affordable and quicker to put in the field. More recently, the ready availability of large troves of online data and big data analysis techniques have opened up all kinds of new research possibilities (Mayer-Schonberger and Cukier, 2013).

There are risks with any methodological innovation. The first is that the legitimate claims that new methods can make to enhancing our understanding of major social questions and augmenting existing research methods become totalizing. The development of survey methods in the middle of the twentieth century was coupled with the creation of a discourse around positivism that equated these methods with being able to uncover innate truths through the deployment of a scientific method. Such accounts neglected the assumptions and shortcomings of the method (Herbst, 1993). Recent hyperbolic claims that big data analysis offers an entirely new epistemological paradigm, where knowledge claims are wholly reliant on correlation and where theoretical questions can safely be neglected, represent the modern version of such discourses (Anderson, 2008). Discourses of this kind matter, not just because they construct the institutional incentives in academia (the possibilities of being published in top journals or winning funding awards), but also because they frame what policy-makers regard as appropriate evidence on which to base major decisions. Therefore, it is vital that political communication researchers, whatever their own methodological inclinations and skills, remain advocates of methodological openness and pluralism.

One very contemporary challenge facing political communication researchers relates to how they obtain the datasets that are required for their studies. This is especially true when the datasets relate to major online platforms. Alphabet, the parent company of Google (founded 1998) is currently (2019 Q2) the fourth-largest company in the world by market capitalization. Facebook (founded 2004) is currently the sixth-largest company in the world. In the space of just a few years, these companies have gone from tiny start-ups to huge global players, arguably more powerful than many countries. How are political communication researchers to co-exist with such corporate behemoths?

The emergence of these firms presents an important challenge for researchers. In the era of mass media, datasets of media content were fairly easy to obtain, as they were publicly available and everybody viewing them was exposed to broadly the same content. In contrast, online media, especially social media platforms, tailor their content heavily to individuals. There is no shared audience experience in the traditional sense. This makes it far harder to gather data, to know who is seeing what, or why they are seeing it.

Early research on social media platforms tended to focus on user-generated content and the sites' social dimension, particularly as it related to self-organization and mobilization. However, the core business model of these firms is advertising, so political communication scholars would naturally be drawn to studying both the content and the effects of such sites. This interest has only intensified as online platforms in general and Facebook in particular became central to political events, with questions being raised about the use of targeted political advertising, and accusations about the circulation of 'fake news'.

However, obtaining data presents a Faustian choice for researchers. One approach involves working with the firms. In response to the events of 2016,

Facebook opened up an initiative to allow researchers to bid for both money and access to their data. The initiative was to be organized by an independent commission (Social Science One, 2019). The idea was that this would remove Facebook from the decision-making process surrounding the research. Thus, the company would have no say in who was granted access to data, nor any access to research findings prior to publication. While commendable, such an approach presents serious challenges. As the datasets used to undertake studies are never publicly released, replication studies are not possible. Furthermore, because researchers have to bid for access to the data, and that access is limited, this model creates a division between the 'data-haves' and the 'data-have-nots' in academia.

An alternative approach is the Open API (Application Programming Interface) model of data collection. This approach has been popular with researchers studying Twitter, and goes a long way to explaining the disproportionate amount of research on that social media platform, certainly relative to the number of users it has, or its value as a company. In contrast to the Facebook model, Open APIs allow researchers to build applications to interact with the platforms, to download massive amounts of data directly from social media platforms, and subsequently control that data in a way that allows them to do anything they wish with it. Such an approach has been compared favourably with the Facebook model by some researchers (Bruns, 2018).[2]

However, there are problems with the Open API approach. First, research-ers will often find it hard to understand how their dataset sits in the wider data universe it is extracted from and the extent to which it is representative (Anstead et al., 2018). Second, the APIs are inherently unstable, existing only because a company has chosen to allow researchers to use them and, as a result, they can be taken down with no notice. Facebook has done exactly this, shut-ting down with no notice some of the APIs that researchers and campaigners have been using to examine Facebook political advertising (Waterson, 2019). The ostensible reason for this shutdown was protecting users' privacy.

This excuse may seem rather self-serving (after all, the applications that used the Facebook API to gather data required volunteers to install browser plug-ins with very clearly written and well-explained terms and conditions, a level of informed consent that Facebook has arguably failed to deliver to its users over the years), but it does highlight the big tension demonstrated by the Cambridge Analytica scandal. On the one hand, Facebook advertising – which is very secretive, with adverts appearing on individual users' timelines through a mechanism that is non-transparent – urgently needs to be opened up to both election regulators and civil society generally (including academics researching the field). Only with access to data can the effects of advertising of this kind and the processes behind it be properly understood, and the necessary regulatory oversight occur. On the other hand, though, the original seed dataset used by Cambridge Analytica was allegedly harvested for an academic project and then – in breach of the licensing agreement, UK data protection law and academic research ethics – used to create a commercial political project (Cadwalladr and Graham-Harrison, 2018a).

Neither the collaboration model of research nor the Open API approach is perfect or can hold all the answers. As a result, the best insights will only come when the two approaches are effectively married. Researchers embedded within the ecosystem of the social media corporations should not dismiss the work of scholars outside that space due to the potential weaknesses of their dataset. Neither should more critical researchers without access to curated social media datasets reject the work of those inside as 'corporate shills'. Instead, they should both be advocating for the value of each other's approach.

Why the direction of political communication research matters: The growing centrality of political communication to politics

While many of the challenges facing contemporary researchers would have been familiar in the past (in form, if perhaps not in precise content), it is also worth noting an additional change in political communication research which has made the stakes much higher. Those of us who spend our lives researching political communication will require little persuasion of the importance of the questions we study. However, taking a step back from our own professional interest in the field, it is genuinely striking how there has been a broadening acknowledgement of the centrality of questions related to political communication to debates about wider political life. This applies both in the wider discipline of academic politics and in public discourse about politics.

In part, this development has been driven by political events in the past two decades. Whether we are seeking to understand Islamic fundamentalist terrorism (radicalization and online networks), subsequent Western military interventions (government communication practices), the Arab Spring (the so-called Twitter revolutions), or the rise of populist political leaders (incendiary political rhetoric, Facebook advertising and 'fake news'), the communication dimension of contemporary political events is inescapable.

A brief survey of top academic politics journals also provides evidence of an increased focus on political communication-related topics. Using a list of the top 15 political science journals, the Scopus database was searched for publications that mentioned 'political communication' or 'media' in their title or abstract, and were published between 1980 and 2018 (the list of top-ranked journals came from Scopus, 2018).[3] The results are shown in Figure 2.1. This is particularly notable in an upward trend that has developed in the years since 2007, with a greatly increased number of articles in top political science journals including references to topics related to political communications and the media.

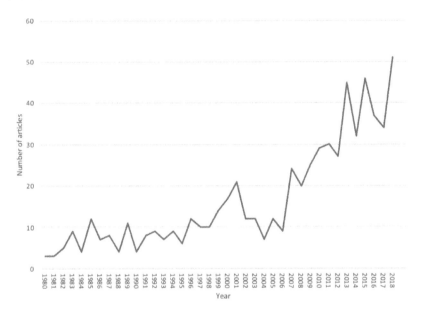

Figure 2.1   Number of articles in leading political science journals mentioning 'political communication' or 'media' in titles or abstracts, 1980–2018 (n = 645)

The increased focus on political communication in academic political science has been mirrored in broader political discourse. A number of major public enquiries have examined issues related to political communication in a variety of countries. This included the judge-led Levenson Inquiry into press regulation in the UK, and the special prosecutor investigation led by Robert Mueller in the United States, which covered questions of Russian interference in the 2016 elections. Mueller's focus included issues related to data leaks and illicit Facebook campaigning. Two recent British prime ministers have used their valedictory speeches to talk about issues related to political communication (Tony Blair on the so-called 'feral' media in 2003; Theresa May on the tenor of political debate in 2019). In such an environment, then, the work of and insights provided by political communication scholars could not be more important.

Conclusion: what should the future of political communication research look like?

This chapter has offered some goals for which political communication researchers should strive. These include: having an interest in new technologies, but a desire to frame our study of them in broader media, political and social systems; being thoughtful about identifying the real challenges (and possible threats) facing liberal democratic institutions; being open to both theories and empirical work developed in a variety of countries and contexts; being methodologically pluralist; and operating both inside and outside the space provided by major media corporations.

It might seem contradictory, but the twin tasks of political communication scholars could be described as inclusion and de-centring. First, we must ensure that the communication angle is not neglected when political questions are addressed. The increased focus on political communication in both general political science research and contemporary political discussions suggests that we are being successful in the goal of inclusion. However, this success makes our second task of de-centring even more urgent. By this, I mean trying to correct overly media-centric explanations for social and political phenomena. Our job is to push back against simplistic narratives which make a narrow aspect of the communication environment (that is, a single social media platform) solely responsible for complex political and social phenomena.

This is also true of the old debate between the so-called Internet pessimists and optimists. In brief, the optimists argued that the Internet has the potential to improve the quality of democratic life, while the pessimists believed that the new technology would undermine democratic institutions. The original advocates of these positions saw them as empirical questions to be tested by empirical methods. However, they neglected the extent to which they could also be discourses, framing debate about new media and political institutions at particular moments in time, driven by the democratic temper of the moment. At certain points in recent history (the immediate aftermath of Barack Obama's 2008 presidential election victory, for example) the optimistic viewpoint dominated. By 2016, however, the pessimistic viewpoint had come to be completely prevalent. The reality is that both can be true simultaneously, and there is nothing contradictory in this.

This challenge is perhaps best understood through the old adage that the job of an academic is to make complicated things simple and simple things complicated. However, this cliché neglects the possibility that these approaches might be required in different measures at different times and in different fields of research. In the current climate, the role of political communication researchers is certainly the latter half of the equation: making things that appear simple and subject to monocausal explanations more complex and nuanced. By injecting such ideas into public debate, the contribution we can make is potentially immeasurable.

Notes

1. This is at least true of the modern form of political communication research. Historians have obviously had a long-running interest in the consequences of the development of the printing press (see, e.g., Eisenstein, 1980).
2. In the interests of transparency, it should be noted that the author is a signatory to Bruns' note.
3. The journals surveyed were based on the 2018 top general political journals in Scopus. Security studies journals were excluded. The journals used were: *American Journal of Political Science, American Political Science Review, Comparative Political Studies, Journal of European Public Policy, The Journal of Politics, JCMS: Journal of Common Market Studies, Journal of Democracy, British Journal of Political Science, Political Studies, Party Politics, Electoral Studies, European Journal of Political Research, Annual Review of Political Science, World Politics* and *West European Politics*. This search generated a total of 645 articles.

References

Anderson, C. (2008) 'The End of Theory: The Data Deluge Makes the Scientific Method Obsolete', *Wired*. Available at https://www.wired.com/2008/06/pb-theory/ (accessed 7 September 2019).

Anstead, N., Magalhães, J.C., Stupart, R. and Tambini, D. (2018) Political Advertising on Facebook: The Case of the 2017 United Kingdom General Election. Paper presented at the European Consortium of Political Research, University of Hamburg, Germany. Available at https://ecpr.eu/Filestore/PaperProposal/71b9e776-0ea8-4bf3-943e-d25fa26898b8.pdf/ (accessed 7 September 2019).

Bruns, A. (2018) 'Facebook Shuts the Gate after the Horse Has Bolted, and Hurts Real Research in the Process'. Available at https://medium.com/@Snurb/facebook -research-data-18662cf2cacb/ (accessed 8 August 2019).

Cadwalladr, C. and Graham-Harrison, E. (2018a) 'How Cambridge Analytica Turned Facebook "Likes" into a Lucrative Political Tool', *Guardian*. Available at https:// www.theguardian.com/technology/2018/mar/17/facebook-cambridge-analytica -kogan-data-algorithm/ (accessed 12 August 2019).

Cadwalladr, C. and Graham-Harrison, E. (2018b) 'Revealed: 50 Million Facebook Profiles Harvested for Cambridge Analytica in Major Data Breach', 17 March. https://www.theguardian.com/news/2018/mar/17/cambridge-analytica-facebook -influence-us-election.

Chadwick, A. (2013) *The Hybrid Media System: Politics and Power*. Oxford: Oxford University Press.

Chadwick, A., Vaccari, C. and O'Loughlin, B. (2018) 'Do Tabloids Poison the Well of Social Media? Explaining Democratically Dysfunctional News Sharing', *New Media and Society*, 20(11), pp. 4255–74.

Eisenstein, E.L. (1980) *The Printing Press as an Agent of Change*. New York: Cambridge University Press.

Garrett, R.K., Bimber, B., De Zúñiga, H.G., Heinderyckx, F., Kelly, J. and Smith, M. (2012) 'New ICTs and the Study of Political Communication', *International Journal of Communication*, 6, p. 18.

Herbst, S. (1993) *Numbered Voices: How Opinion Polling Has Shaped American Politics*. Chicago, IL: University of Chicago Press.

Mayer-Schonberger, V. and Cukier, K. (2013) *Big Data: A Revolution That Will Transform How We Live Work and Think*. London: John Murray.

Murgia, M., Findlay, S. and Schipani, A. (2019) 'India: The WhatsApp Election', *Financial Times*. Available at https://www.ft.com/content/9fe88fba-6c0d-11e9-a9a5 -351eeaef6d84/ (accessed 7 September 2019).

Putnam, R.D. (2001) *Bowling Alone: The Collapse and Revival of American Community*. New York: Simon & Schuster.

Robinson, M.J. (1976) 'Public Affairs Television and the Growth of Political Malaise: The Case of "The Selling of the Pentagon"', *American Political Science Review*, 70(2), pp. 409–32.

Runciman, D. (2017) *The Confidence Trap: A History of Democracy in Crisis from World War I to the Present*, rev. edn. Princeton, NJ: Princeton University Press.

Scopus (2018) 'Scimago Journal and Country Rank: Political Science'. Available at   https://www.scimagojr.com/journalrank.php?category=3312&country=GB/ (accessed 12 August 2019).

Social Science One (2019) 'Social Science One: Building Industry–Academic Partnerships'. Available at https://socialscience.one/ (accessed 12 August 2019).

Waisbord, S. (2019) *Communication: A Post Discipline*. Hoboken, NJ: Wiley.

Waterson, J. (2019) 'Facebook Restricts Campaigners' Ability To Check Ads for Political Transparency', *Guardian*. Available at https://www.theguardian.com/ technology/2019/jan/27/facebook-restricts-campaigners-ability-to-check-ads-for -political-transparency/ (accessed 18 February 2019).

Wolfsfeld, G., Segev, E. and Sheafer, T. (2013) 'Social Media and the Arab Spring: Politics Comes First', *International Journal of Press/Politics*, 18(2), pp. 115–37. doi: 10.1177/1940161212471716.

3 As it was in analogue days: the relevance of legacy research

Jay G. Blumler

Introduction

Political communication processes, their possible outcomes, and the need for research into them have undeniably been transformed by digitization. Much of the model of political communication systems prevalent in the practice and scholarship of liberal democracies in the twentieth century has become out of date in many respects. What was once a pyramidal, top-down system has become an up-and-down and reciprocally round-the-houses one. The ever-expanding diffusion and utilization of Internet facilities and social media platforms has unleashed an incredibly diverse range of globally expansive and temporally synchronous communicative networks. New flows and qualities of interpersonal political discourse are in play. From being predominantly receivers of institutionally organized communications, members of what were once called 'the audience' have become a communicative force – or set of forces – in their own right. An ecology of two different levels of political communication – institutionalized and grassroots – has emerged.

Why analogue-period research is not obsolete

There is a danger that younger scholars entering the field in this period, and keen to build impressive research and publication records rapidly may treat it like a *tabula rasa*, on which to inscribe their contributions in digital bits and pieces. Such a tendency could be bolstered by certain other features of our subject:

- that political communication has always been an eclectic field, reliant to some extent on theories produced in other disciplines, such as political science, sociology, journalism studies and electoral studies;
- the increasing diversification and specialization across the board of all branches of communication study;
- the wide variety and multiplicity of topics available for political communication research attention, more so in digital circumstances, of course (see the Contents page of this volume).

But, while opening the new digital book, the old analogue one should also be kept open by political communication scholars, for several reasons. Firstly, some important features of political communication will very likely abide. Incentives to attract large audiences for messages and appeals through mainstream media will still be operative even if alongside other drives. The roles of specialist political consultants in devising strategies for the conduct of communication campaigns, which came forth in the 1980s, could be even more crucial for political actors in today's highly complex environment. Secondly, a wealth of theories and empirically supported findings have emerged from legacy-period research that should be built on, not sidelined. Thirdly, past research is like a laboratory, in which various ways of tackling problems of enquiry have been fruitfully tried and disclosed. Fourthly, the pioneering leaders of the field strove for heights and forms of creativity with which present-day scholars should become familiar and where possible aim to emulate.

Thinking big

What features and specimens of the analogue corpus should digital scholars be encouraged to consider? To answer that question, I rummaged through several collections of studies, which their editors regarded as 'the most durable work' in mass communication research published in the previous year. Specially illuminating were a set of six Mass Communication Research Yearbooks (MCRY), published by SAGE from 1980 to 1987 under editorial teams at Indiana, Illinois and Maryland universities, respectively.

Two more general characteristics of analogue-period work and thought stood out from that perusal. Firstly, it was evident that in the latter half of the twentieth century, mass and political communication scholars often aimed to sketch out what might be termed a 'big picture' (or big and alternative and opposing pictures) in their areas of concern. It was as if reflection was a *sine qua non* over and above the conscientious reporting of discrete results. An editor of Volume 2 of the MCRY, for example, noted a great concern over the direction and depth of mass communication research among many of the contributors. Indeed, large chunks of most of the Yearbooks were devoted to 'theoretical perspectives', and broad and summary treatments were often given to what could be said or be supposed about a host of matters and not just over issues in the debate between critical theorists and liberal democratic pluralists. Thus, essays traded in large-scale propositions on the functions of the mass media for society; media relations to social and political change and control; the sources and varieties of media effects; contrasts of media realities with 'actual' realities; media roles in political socialization; transmission versus cultural models of communication; the interplay of engagement in interpersonal communications and exposure to mass communications; the impact of critical events on public opinion; and elite pluralist versus participatory notions of democracy. Gladys Lang even urged colleagues in 1987 to explore 'the role of mass communication in the survival of the world'. The production of grander perspectives may be more difficult, but it is also more necessary, to make coherent sense of political communications in our centrifugally fragmented societies.

Secondly, I was struck by the ways in which empirical research and its results were often introduced, organized and presented. Surprisingly absent was the model for this that tends to dominate journal articles nowadays: literature review; specification of something not yet or imperfectly addressed; formation of hypotheses; methods for testing them; data and measures deployed; findings; implications for the hypotheses; conclusion. Instead, a typical point of departure for many authors was a set of assumptions about a phenomenon, an enquiry approach, or a widely accepted theory, the adequacy of which should in their view be properly researched. For example, Rothenbuhler (1987) set out to probe the usefulness of six elements of neofunctionalist theory for analysing the relationship between mass communication and other subsystems of society and politics. Rowland (1986) kicked off by mentioning 'the primary assumptions on which the communications policy literature has dealt with problems of technological change'. Chaffee and Hochheimer (1982) began by identifying 'four basic assumptions about "what we know" [supposedly] about the importance of mass media'. Cook et al. (1983) entitled their article about the impacts of TV violence on behaviour, 'The Implicit Assumptions of Television Research'. Hackett (1984) referred to 'four assumptions on which conventional research on media bias is based'. At the outset of his piece, Babe

(1983) stated, 'This article addresses some of the major assumptions and modes of analysis of traditional microeconomic theory.'

Of course, both strategies can be advantageous. By facilitating a cumulation of findings, the dominant model may help to serve the aim of developing a science of communications. But the continuing validity of its propositions is at least questionable given the ever-changing dynamics of media organizations and socio-political systems, and the immense variety of systems present in different parts of the world. More insidiously, it may impose straightjackets on creativity. The analogue scholars' approach was probably more amenable to the generation of fresh ideas and unexpected findings. A case can be made, not to abandon the current model, but to avoid socializing and training budding scholars to regard it exclusively as the only way forward for them to adopt.

A curricular proposal arises from all this. To acquaint PhD students with what analogue scholarship might have to offer, why not give them a seminar course based on the reading of a set of outstanding publications, the potential con-temporary research relevance (or irrelevance) of which they would be invited to discuss? My own list of classics that could be used for this purpose follows (see Box 3.1), but course leaders might prefer to assign other analogue-period readings.

Box 3.1   Analogue-period readings on political communication

Lazarsfeld and Merton (1948) 'Mass Communication, Popular Taste and Organized Social Action'.
Westley and MacLean (1957) 'A Conceptual Model for Communications Research'.
Tichenor et al. (1970) 'Mass Media Flow and Differential Growth in Knowledge'.
McCombs and Shaw (1972) 'The Agenda-Setting Function of the Media'.
Katz, Blumler and Gurevitch (1973) 'Uses and Gratifications Research'.
Kraus et al. (1975) 'Critical Events Analysis'.
Robinson (1976) 'Public Affairs Television and the Growth of Political Malaise'.
Adoni (1979) 'The Functions of Mass Media in the Political Socialization of Adolescents'.
Blumler and Gurevitch (1981) 'Politicians and the Press: An Essay on Role Relationships'.
Lowenthal (1984) 'The Triumph of Mass Idols'.

Swanson (1992) 'Managing Theoretical Diversity in Cross-National Studies of Political Communication'.
Schulz (2014) 'Political Communication in Long-Term Perspective'.

Past work and future research agendas

I contend that the generation of future research agendas could be facilitated by revisiting the following legacy-period theories.

Uses and gratifications

Some current scholars have adopted this approach out of a natural interest in studying people's responses to the onset of a host of new communication technologies, platforms and devices. The focus of much of this work, however, has been individuals' motives for engaging with digital material and the gratifications they have enjoyed when doing so – full stop. This fails to take account of the much more extended original formulation of this approach, comprising in Katz et al.'s (1973) words, 'the social and psychological origins of needs, which generate expectations of the mass media or other sources, which lead to differential patterns of media exposure (or engagement in other activities), resulting in need gratifications and some other consequences, mostly unintended ones'. A richer agenda could be devised by building some of these other elements into gratifications research on new media, taking on board some of the issues raised about the approach in the 1970s and 1980s concerning, for example, the extent and forms of audience activity, the relationship between gratifications sought and obtained, relations to functional alternatives, and the implications for communication effects.

Journalists' roles

An impressive body of concepts and findings about these, both as ideally professed and as actually practised on the job, has been produced on a global scale but chiefly for journalists working in mainstream enterprises in the organizational, financial, competitive and cultural conditions that apply to them. But, as Hanitzsch pointed out in 2007, 'journalism culture is a fast-changing object of enquiry'. Future research should be undertaken on these roles as conceived and acted upon, both within mainstream organizations striving to adapt to the digital environment and in the various start-ups that are trying to do journalism differently. One aim of such work could be to consider for possible revision or supplementation the typology of role orientations, both as notions and as practised, that emerged from the World of Journalism Study's transnational interviews of journalists in 18 countries (Hanitzsch, 2011):

- detached watchdog – for example, to expose abuses of power;
- critical change agent – for example, to promote national development;
- populist disseminator – for example, to cater for audiences' interests;
- opportunistic facilitator – for example, to support a government's or a political party's goals.

Attention to news and politics

In past research, a distinction between those individuals who deliberately choose to follow political news and propaganda and those who have been incidentally exposed to such material has repeatedly been found to be empirically significant. Media effects have tended to be concentrated among the latter; the former have been less readily persuasible and more likely to be reinforced in their initial opinions. But with the proliferation and diversification of ways in which political information can now reach people, the question arises of whether the analogue pattern still survives, and for whom it belongs on the digital research agenda. Is incidental exposure more common than it used to be? If so, has it increased news comprehension or ignorance? Does it still have a persuasion effect potential? Or are such consequences likely to be smothered by the rapid-fire bombardment of many different domestic and international news items and rhetorical ploys that increasingly hit citizens willy-nilly? And what about the apparently increasing group of news avoiders? Might its emergence require some revision of the pattern that analogue researchers postulated? Uses and gratifications research could play a part in the development of this agenda. In a host of studies, the most frequently endorsed reason for following news and current affairs has been surveillance of the political environment (for example, to learn what politicians might do if given power, to judge what political leaders are like, to learn more about the issues of the day). Is that still the case?

Political communication change

Trends over time

Political communication is eminently a dynamic process, ever in flux. Such change can be incremental, or happen in major bursts. If such change must be on scholars' agendas, Winfried Schulz's (2014) essay, 'Political Communication in Long-Term Perspective', is a 'must read' for anyone wishing to gain a serious grip on it. In a magisterial overview of how political communication change was theorized and measured in analogue days and is being approached in digital ones, Schulz summarizes the literature through three perspectives. One taps change at two points in time, an approach that can be relevant for landmark political communication theories, such as the spiral of silence, the knowledge gap hypothesis, agenda-setting, framing and the diffusion of innovations. A second approach has used concepts ending in 'ization', such as professionalization, mediatization and tabloidization, and latterly, fragmentation, embedded by some authors in analyses of social or political change, such as modernization. And thirdly, there are phase models, exemplified by Blumler and Kavanagh's (1999) analysis of three postwar 'ages' of political communication. Schulz concludes by commenting on media roles in change perspectives, especially that of television in the legacy period, as well as its interplay with the Internet in the twenty-first century. The latter has prompted diverse normative positions, with some authors bemoaning how political communication seems to be 'on the slide', while others celebrate apparent enhancements of individual citizens' autonomy and information resources. Both optimistic and pessimistic views seem to be apposite.

Critical events analysis

Every now and then, a major political event may disrupt ongoing processes, challenge and galvanize leading political actors, and receive prominent and dramatic coverage in the media. Writing in 1975, Kraus et al. urged political communication scholars to give agenda priority to them, and outlined how they could be fruitfully researched. As they put it, 'Critical events analysis seeks to identify those events which will produce the most useful explanations . . . of social change'. This, however, should not just be about the phenomena per se, because 'the social conditions which reinforce or nullify the impact of events must [also] be specified'. For this purpose, 'attempts' must be made 'to integrate data collected at the individual and societal levels of analysis'. Advocating a rounded approach, critical events analysis would examine, they say, the extent and modes of elite responses, the intentions and efforts of journalists, the kinds of messages that ensue, the 'public's attention to and use of messages about dramatic events', and how the event comes to be portrayed in the media and in citizens' understandings. An outstanding empirical example – indeed a highly recommendable model – of this approach can be found in Lang and Lang's (1983) book-length study of all that was involved in the Watergate scandal, entitled *The Battle for Public Opinion*. Political communication scholars should be on the look-out for the occurrence of events that could be amenable to equivalent treatment.

Campaign communication strategies

The literature about this in both analogue and digital circumstances appears quite full and is undoubtedly worthy of continual pursuit. In the creative spirit of the field's pioneers, however, there is a case for dimensionalizing afresh how communication strategies, as designed by consultants and implemented by political actors, are conceived and researched (ethnographically, in content analyses and elite interviews). A starting point (normative, admittedly) could be a recognition that communication strategies may vary in their potential for civic enhancement. Some among the following dimensions might be posited and researched:

- what is assumed about the appetite of citizens to learn what politicians might do if given power;
- level of policy emphasis, that is, whether positions are simply asserted or are amplified with justifications, explanations and testable evidence;
- kinds of news values catered for;
- whether rivals are regarded as opponents or as enemies;
- whether public dissatisfaction with the prevailing system is exploited for advantage or regarded as something that should be countered;
- fidelity to ethical scruples or not;
- whether mass perceptions of what is going on in politics are deemed more important than substantive realities;
- degree of willingness to be questioned by journalists and citizens, versus attempted close control of appearances in the media and on the ground.

Comparative analysis of political communication

Theory in this area has been dominated by Hallin and Mancini's (2004, 2012) typology of three models of media systems: liberal, polarized pluralist and democratic corporatist. But these apply to mainstream media, and the impact on them of the onset of digital communications, the Internet and social media, little studied so far, should be put more prominently on the field's agenda. Key questions here are whether the models are becoming less distinct, more mixed and less differentiated from each other.

Any shortcomings?

To balance its scorecard, some of the weaknesses and blind spots of legacy political communication research should be noted. For one thing, political communication scholars were actually quite slow to 'go comparative'. In 1975, Blumler and Gurevitch described comparative political communication research as 'a field in its infancy' and 'the least advanced topic dealt with' in Chaffee's volume on political communication. And Gurevitch and Blumler declared in 1990 that its 'patchy fruits' had still only progressed to a 'late adolescence'. Political communication scholars were also slow to recognize the potential for citizens to become producers as well as receivers of content. The 'active audience', much touted in the 1980s, was more of an interpreter than a maker of messages.

More fundamentally, few analogue scholars tried to understand how technological change reshapes political communication, despite having fully documented the massive intervention of television into most facets of the political communication process. Typically, technological factors were directly addressed when a particular innovation came on the scene, such as videotex (Weaver, 1983), the videorecorder (Levy and Fink, 1984), the arrival of colour television in place of black-and-white (Winston, 1987), and cable- and satellite-based multiplication of TV channels (Katz, 1996). No attempt fully to analyse how technological developments can impact upon political communication, in conjunction with other social, political and media factors, appears in the literature until Keane (2013).

Analogue scholars' handling of the inherently normative implications of research findings for citizenship and democracy stands out as an even more fundamental weakness. Too often, those implications were briefly tossed off like afterthoughts in an article's concluding passage. Too often, they pivoted on a narrowly taken-for-granted notion of citizenship in a liberal democracy and on how the media do or do not provide information to serve that. A mushrooming of other normative concepts, such as authenticity, transparency, deliberation and inclusiveness, was a much later development. Too rarely did authors explicitly present civic ideas as norms to be debated, distinct from evidence-based matters of fact. No truly full attempt to explore all that a normative umbrella for politics and mediated communication can involve appears in the literature until Christians et al. (2009).

Conclusion

My concern that analogue-period research might be neglected by present-day scholars may or may not be misplaced. I shall therefore be very interested to discover whether other contributors to this book have taken account of 'how it was in analogue days' and, if so, how.

References

Adoni, H. (1979) 'The Functions of Mass Media in the Political Socialization of Adolescents', *Communication Research*, 6(1), pp. 84–106.

Babe, R.E. (1983) 'Information Industries and Economic Analysis: Policy Makers Beware'. In O.H. Gandy Jr, P. Espinosa and J.A. Ordover (eds), *Proceedings from the Tenth Annual Telecommunications Policy Research Conference*. Norwood: Ablex, pp. 123–35.

Blumler, J.G. and Gurevitch, M. (1975) 'Towards a Comparative Framework for Political Communication Research'. In S.H. Chaffee (ed.), *Political Communication: Issues and Strategies for Research*. Beverly Hills, CA: SAGE, pp. 165–93.

Blumler, J.G. and Gurevitch, M. (1981) 'Politicians and the Press: An Essay on Role Relationships'. In D.D. Nimmo and K.R. Sanders (eds), *Handbook of Political Communication*. Beverly Hills, CA: SAGE, pp. 467–93.

Blumler, J.G. and Kavanagh, D. (1999) 'The Third Age of Political Communication: Influences and Features', *Political Communication*, 16(3), pp. 209–30.

Chaffee, S.H. and Hochheimer, J.L. (1982) 'The Beginnings of Political Communication Research in the United States: Origins of the "Limited Effects" Model'. In E.M. Rogers and F. Balle (eds), *The Media Revolution in America and Western Europe*. Norwood: Ablex, pp. 267–96.

Christians, C.G., Glasser, T.L, McQuail, D., Nordenstreng, K. and White, R.A. (2009) *Normative Theories of the Media: Journalism in Democratic Societies*. Urbana and Chicago, IL: University of Illinois Press.

Cook, T.D., Kendzierski, D.A. and Thomas, S.V. (1983) 'The Implicit Assumptions of Television Research: An Analysis of the1982 NIMH Report on Television and Behavior', *Public Opinion Quarterly*, 47(2), pp. 161–201.

Gurevitch, M. and Blumler, J.G. (1990) 'Comparative Research: The Extending Frontier'. In D.L. Swanson and D. Nimmo (eds), *New Directions in Political Communication: A Resource Book*. Newbury Park, CA, USA; London, UK; New Delhi, India: SAGE, pp. 305–25.

Hackett, R.A. (1984) 'Decline of a Paradigm? Bias and Objectivity in News Media Studies', *Critical Studies in Mass Communication*, 1(3), pp. 229–59.

Hallin, D.C. and Mancini, P. (2004) *Comparing Media Systems: Three Models of Media and Politics*. Cambridge, UK and New York, USA: Cambridge University Press.

Hallin, D.C. and Mancini, P. (eds) (2012) *Comparing Media Systems beyond the Western World*. Cambridge, UK and New York, USA: Cambridge University Press.

Hanitzsch, T. (2007) 'Deconstructing Journalism Culture: Toward a Universal Theory', *Communication Theory*, 17(4), pp. 367–85.

Hanitzsch, T. (2011) 'Populist Disseminators, Detached Watchdogs, Critical Change Agents and Opportunistic Facilitators: Professional Milieus, the Journalistic Field and Autonomy', *International Communication Gazette*, 73(6), pp. 477–94.

Katz, E. (1996) 'And Deliver Us from Segmentation', *Annals of the American Academy of Political and Social Science*, 546(1), pp. 22–33.

Katz, E., Blumler, J.G. and Gurevitch, M. (1973) 'Uses and Gratifications Research', *Public Opinion Quarterly*, 27(4), pp. 509–23.

Keane, J. (2013) *Democracy and Media Decadence*. Cambridge: Cambridge University Press.

Kraus, S., Davis, D., Lang, G.E. and Lang, K. (1975) 'Critical Events Analysis'. In S.H. Chaffee (ed.), *Political Communication: Issues and Strategies for Research*. Beverly Hills, CA, USA and London, UK: SAGE, pp. 196–216.

Lang, G.E. (1987) 'The Personal and the Public: A Very Personal Public Response'. In M. Gurevitch and M.R. Levy (eds), *Mass Communication Review Yearbook*, Vol. VI. Beverly Hills, CA: SAGE Publishing, pp. 39–42.

Lang, G.E. and Lang, K. (1983) *The Battle for Public Opinion: The President, the Press and the Polls during Watergate*. New York: Columbia University Press.

Lazarsfeld, P.F. and Merton, R.K. (1948) 'Mass Communication, Popular Taste and Organized Social Action'. In L. Bryson (ed.), *The Communication of Ideas*. New York: Harper & Brothers, pp. 95–108.

Levy, M.R. and Fink, E.L. (1984) 'Home Videorecorders and the Transience of Television Broadcasts'. In M. Gurevitch and M.R. Levy (eds), *Mass Communication Review Yearbook*. Beverly Hills, CA, USA; London, UK; New Delhi, India: SAGE, pp. 56–71.

Lowenthal, L. (1984) 'The Triumph of Mass Idols'. In L. Lowenthal (ed.), *Literature and Mass Culture*. New Brunswick, NJ, USA and London, UK: Transaction Books, pp. 203–35.

McCombs, M. and Shaw, D. (1972) 'The Agenda-Setting Function of the Media', *Public Opinion Quarterly*, 36(2), pp. 176–87.

Robinson, M.J. (1976) 'Public Affairs Television and the Growth of Political Malaise: The Case of the Selling of the Pentagon', *American Political Science Review*, 70(2), pp. 409–32.

Rothenbuhler, E.W. (1987) 'Neofunctionalism for Mass Communication Theory'. In M. Gurevitch and M. Levy (eds), *Mass Communication Review Yearbook*, Vol. 6. Beverly Hills, CA, USA; London, UK; New Delhi, India: SAGE, pp. 67–85.

Rowland, W.D. (1986) 'American Telecommunications Policy Research: Its Contradictory Origins and Influences', *Media, Culture and Society*, 8(2), pp. 159–82.

Schulz, Winfried (2014) 'Political Communication in Long-Term Perspective'. In C. Reinemann (ed.), *Political Communication*. Berlin: De Greyter Mouton. pp. 63–85.

Swanson, D.L. (1992) 'Managing Theoretical Diversity in Cross-National Studies of Political Communication'. In J.G. Blumler, J.M. McLeod and K.E. Rosengren (eds), *Comparatively Speaking: Communication and Culture across Space and Time*. Newbury Park, CA, USA; London, UK; New Delhi, India: SAGE, pp. 19–34.

Tichenor, P.J., Donohue, P.A. and Olien, C.N. (1970) 'Mass Media Flow and Differential Growth in Knowledge', *Public Opinion Quarterly*, 34(2), pp. 159–70.

Weaver, D.H. (1983) 'Teletext and Viewdata'. In M. Gurevitch and M.R. Levy (eds), *Mass Communication Review Yearbook*. Beverly Hills, CA, USA; London, UK; New Delhi, India: SAGE, pp. 581–607.

Westley, B.H. and MacLean, M.S. (1957) 'A Conceptual Model for Communications Research', *Journalism Quarterly*, 34(1), pp. 31–8.

Winston, B. (1987) 'A Whole Technology of Dyeing: A Note on Ideology and the Apparatus of the Chromatic Moving Image'. In E. Wartella and D.C. Whitney (eds), *Mass Communication Review Yearbook*, Vol. 4. Newbury Park and Beverly Hills, CA, USA; New Delhi, India: SAGE.

PART II

Campaigns and Elections

4 Political parties in the digital era

Paolo Gerbaudo

Introduction

It has become a bit of a techno-utopian stereotype to speak of a digital revolution as a sort of tsunami upsetting all aspects of our life (Morozov, 2012). However, it is fair to say that to some extent this discourse, which was for a long time promoted by *Wired* magazine and other well-known voices of California techno-utopia, seems to have eventually come to fruition in all spheres. This discourse included the political sphere, forecasting various forms of consumption, social relationships, commerce, and more recently also political processes, being radically transformed by the irruption of digital technology.

We live in a society in which many fundamental processes, and in particular processes of communication, have little resemblance to the reality of a few decades ago. However, this transformation has not proceeded along those unambiguously positive lines that the techno-prophets had predicted. We are in a world that has been radically transformed, but in which this transformation, contrary to what was promised 'on the tin' carries at least as many downsides as it carries upsides. What is more, we live in a world that still, by and large, we fail to fully comprehend, as we often use old categories to explore new phenomena, to the point of sometimes despairing and using no categories at all, falling prey to merely descriptive analysis.

This highly contradictory and ambiguous terrain is particularly the case when it comes to the study of political parties in the digital era. The transformation of politics in the digital era, as it is manifested in all sorts of phenomena, from social movements to political parties, new forms of political communication and social media, has in fact been the object of growing attention in recent years, both from academia and the news media, and among political practitioners themselves. A number of phenomena have made this research agenda particularly urgent.

One has been the explosion of the 2011 protest wave, from the Arab Spring to the Indignados and Occupy Wall Street, which has attracted widespread attention to the role played by social media as a new space of communication, with consequences for organization and mobilization (Gerbaudo, 2012; Tufekci, 2017). It has led to the coining of a number of new catchphrases, such as Facebook or Twitter 'revolutions', and an intense debate has developed over the nature of these transformations.

More recently, many have commented on the connections between the rise of right-wing populism and the new social media-powered public sphere, and on the implications of such trends as fake news and trolling. Much has been said about the transformation of political parties in the digital era and about their internal organizational structures (Margetts, 2001; Norris, 2001; Deseriis, 2017; Gerbaudo, 2019). The emergence of new formations, sometimes described as digital parties, including the Pirate Parties in Scandinavian countries, the Five Star Movement in Italy, Podemos in Spain and many other formations (Box 4.1), has sparked a discussion about the transformation of political parties in the digital era. These developments have been faced with major conceptual difficulties and analytical confusions. The number of publications in this area has ballooned, and has run into the thousands of journal articles, books and book chapters year by year. Yet, we are far from possessing a comprehensive research agenda, not to speak of a common analytical framework and methodology to make sense of these developments.

Box 4.1   New formations: digital parties

Pirate Parties: the first one was founded in Sweden in 2006, and they have since spread all over Northern Europe, from Iceland to Germany. Despite limited electoral success, they have been very influential in introducing new organizational and democratic practices.

Five Star Movement: founded in 2009 by comedian and blogger Beppe Grillo, the Five Star Movement presented itself at the start as a party of the Internet. Ten years since its foundation it is the largest Italian party and it is in government in alliance with Centre-Left Partito Democratico after one year in coalition with right-wing Lega.

Podemos: established in 2014 by Pablo Iglesias and a group of activists and researchers of the Universidad Complutense in Madrid, Podemos has quickly gained electoral backing and heavily used online referendums and similar forms of online participation.

The aim of this chapter is to work towards this goal of sketching out a research agenda for the study of digital politics, with particular reference to the transformation of organizational processes and the emergence of new digital parties. I draw on my experience studying social movements and political parties in the digital era and examine the main research problems that we are currently facing in this area, considering the principal research questions open for discussion. I begin by providing a brief summary of present research at the intersection of digital media and organizational processes. I continue by highlighting the main research questions that are being discussed in this field, highlighting those that are particularly challenging and as yet unsolved.

Parties transforming

Of all the spheres of our society, electoral politics and political parties have long appeared to be the most impervious to political change. Yet, in recent years, a number of phenomena have highlighted how political parties are also progressively transforming, and adapting themselves to the new characteristics of the digital environment. The rise of a number of new political formations (see Box 4.1), such as the Five Star Movement in Italy, Podemos in Spain and Pirate Parties in Scandinavian countries, seems to point to a revival of political parties, and the transformation of political parties to meet the new challenges of the political era. This trend is quite surprising, as it seems to contradict the main thesis that has been dominant in political science over the last two to three decades, namely the assertion of the decline of the political party. Scholars such as Russell J. Dalton (2002), Peter Mair (2013) and many others have in fact argued that parties were losing their central role as mediators between society and political institutions.

Amidst the increasing personalization of politics, the technocratic and post-democratic transformation of government, and the convergence to the centre, political parties seemed to have become pale imitations of their early manifestations. And indeed, it was during this time that political parties experienced a spectacular collapse in membership, with mass parties progressively giving way to electoral-professional parties (Panebianco, 1988). Moreover, the Internet and digital media seemed only to militate against political parties. Their emphasis on the individual and personal networks, and their embedded criticism of all vertical structures as, for example, argued by Manuel Castells (2004) in his influential work on the network society, seemed to point towards a situation in which political parties would have little importance. Yet, the very emergence of new political parties, and of political parties that integrate digital media in their very functioning, seems to point to a very different outcome. Rather than disappearing, parties are morphing into a digital form.

What is required, then, is an understanding of what kind of organization parties are becoming in the present digital era. This is in line with the reasoning of United States scholar David Karpf (2012), who argues that it is useful to distinguish between 'legacy organizations', namely organizations that have been founded before the digital era and are trying to adapt to it, and 'netroots organizations', formations that have emerged in recent years and have consequently been shaped from the very start by digital technology and connected organizational forms.

What matters in this context is that far from being necessarily in contradiction with political organizations, as argued for example by techno-futurist thinkers such as the American pundit Clay Shirky (2008: 67) in his assertion that we are moving to a society of 'organizing without organizations', the Internet has rather favoured the development of new forms of Internet-specific organizations. Collectivity and organization are thus far from having been eliminated. To the contrary, they remain at the centre of any serious research agenda in sociology and political science.

In my own work on political parties (Gerbaudo, 2019), I have argued that formations such as the Five Star Movement, Podemos and Pirate Parties are the manifestation of a new, ideal type of political party, which I described as the 'digital party'. This may sound a rather confounding term at a time of ubiquity of digital media; at this point in time, all parties are, in some sense, digital parties. However, I propose that new parties are digital in a qualified sense. They integrate more fully the Internet and digital media in their internal functioning compared with more traditional or legacy organizations, to use the language of David Karpf (2012). In older organizations, such as traditional political parties, the use of digital technology tends to concern intra-organizational processes and the external communication of parties to their targeted publics.

These organizations tend to be very prudent in absorbing digital technology in their operations, and continue to see television and the press as their main grounds of campaigning. In netroots organizations, instead, the use of digital technology directly concerns the ways in which parties are organized internally and the forms of 'intra-party democracy' (Van Biezen and Piccio, 2013) through which decisions are made. In other words, in traditional parties the transformation wrought by political technologies happens only in their relationship with the outside world; in the case of digital parties proper, the entire life of the party is cast open and rearranged around the idea of a more direct and participatory democracy.

There are a number of key features that distinguish digital parties from more traditional formations. What defines the digital party as a new party type is not simply the embracing of digital technology, but the purpose of democratization which digital technology is called on to fulfil. Organizations such as Pirate Parties, the Five Star Movement and Podemos have presented their adoption of the logic of interactivity and participation, popularized by social media platforms, as a way to deliver a more direct democracy; a democracy which is sometimes imagined to be as smooth as the interactivity of social networking sites and as malleable as the data clouds these services rely upon.

The 'platformization' (Helmond, 2015) of the party, namely its absorption of the logic of digital platforms, is a process that comprises many facets. It is a tendency premised on more or less evident strategic considerations. That is, the platform logic is seen to be more effective and more fitting to present times than the old and bulky bureaucracy of traditional parties. Choosing this path, digital parties attempt to offset their weaknesses as outsider organizations, their lack of steady funding and of offices and similar infrastructures, and their competitive disadvantage vis-à-vis large and well-established organizations.

While this organizational restructuring is evidently efficacy-driven, it cannot be understood as stemming merely from strategic and economic considerations, as a managerial translation to the political arena of the 'lean management' and the 'disruptive innovation' philosophy adopted by several Silicon Valley firms. Rather, this shift is also premised on, and justified by, the utopian vision of an online democracy, using digital technology as a means to extend and deepen political participation, reintegrate in the polity many citizens who have for a long time been distant from the political arena, and allow them to have a more direct and meaningful intervention in the political process.

This research compounds work that has been conducted in recent years by a number of researchers working on similar topics. Helen Margetts (2001), for example, has coined the notion of 'cyber parties', to discuss the transformation of political parties in the digital era. Similarly, Florian Hartleb (2013: 15) has discussed the emergence of 'anti-elitist cyber-parties', to capture the new logic of political parties such as Podemos and the Five Star Movement, which I also discuss in my book (Gerbaudo, 2019). The overarching aim of my own work has been to systematize the common trends that emerge across very different formations in Europe and beyond, and the sometimes contradictory tendencies that underpin them. These include such trends as these parties' anti-bureaucratic spirit and their promise of greater participation for ordinary members on the one hand, and their power centralization and personalized leadership on the other hand.

This trend of transformation within political parties is highly urgent and deserves systematic research, especially given that some of the digital parties have now entered government, as is the case in Italy, while also more traditional and mainstream parties are trying to adapt themselves to this situation and to transform themselves into digital parties using social media and adopting digital democracy practices (Kreiss, 2015). Examples include the Spanish PSOE that has used a digital platform for making decisions on the leadership, or the United Kingdom's Labour Party and its annex organization Momentum that are trying to introduce similar participatory platforms in order to allow their members to participate in discussions. More generally, the use of social media management platforms such as NationBuilder and the restyling of parties as social media-driven machines has become a strong trend across the United States and the United Kingdom, calling for a systematic inquiry.

Faced with this transformation, it is important to avoid the tendency to jump to immediate moral evaluations, often simply brushing over all these processes in the dye of a digital dystopian turn, or as a betrayal of the noble politics of old. Rather, the more difficult and delicate task is to delineate the new forms of power, organization and participation that are emerging in this field, and how they bespeak some more general tendencies of our society. It is only if we avoid a preconceived and moralizing attitude at the start, and acknowledge the novelty and originality of these practices, that we may be in a position to make qualified judgements at the end.

Questions for research

There are two fundamental questions that should be examined by researchers in the coming years in regard to the emergence of digital parties and the turning of all political parties towards this model. These questions are topologically located at the organizational extremes, at the bottom and at the top of the organizational structure of political parties. On the one hand, we are called to discuss the new forms of participation that are emerging in our era, and to engage with notions of membership, forms of involvement, and issues of partisanship, identity and allegiance. On the other hand, there is much to be discussed about the new forms of leadership that are emerging in parties in the digital era, and connected questions of organization, mobilization and strategic thinking that are displayed in contemporary parties.

On the first question, the transformation of political participation in political parties is known to be a highly significant trend. Digital media, and social media in particular, have revolutionized the ways in which people participate in politics. But we have only limited knowledge about these issues. We know that members of contemporary political parties are rather different from the rank-and-file of traditional mass parties, and that they often do not correspond to the expectations of strong encadrement, ideology, and so on, that we attribute to them (see, e.g., Natale and Ballatore, 2014). Research has shown that contemporary forms of participation tend to be more individualized and often more unstable than they were at the height of the industrial era (Bennett and Segerberg, 2012; Chadwick and Stromer-Galley, 2016). Citizens are behaving more like consumers, who can quickly shift from one 'product' to the next, with significantly low levels of loyalty to formations. This is a thesis that was raised earlier by Katz and Mair (1995) in their theory of the cartel party, and before them by Otto Kircheimer (1966) in his discussion of the catch-all party.

Digital parties seem, if anything, to push this logic to the extreme. The ease of access to these parties, which is a consequence of the low barrier to membership, in fact carries the consequence of making them also easier to exit, with evident risks for the effectiveness of these organizations. This contrasts with the state of more traditional parties which had a more stable form of membership and identity. On this question there is still much research that needs to be conducted in order to clarify the new nature of participation and assess the degree to which we are really witnessing a radical abandonment of the prior logics of participation. Furthermore, it is unclear the degree to which this individualization of participation is really a general phenomenon, common to other formations and more traditional organizations as they adapt to the digital era, or whether it only affects the new digital parties. It is therefore necessary to explore the differentiation in participation, identity and membership that exists across different formations, and the ways in which it can be controlled and governed.

Regarding the question of leadership, this is an issue that also requires significant analytical, theoretical and methodological innovations. Much theorizing on the Internet has in fact purported that digital media have an horizontalist nature that would end up eroding hierarchy, organization and, ultimately, leadership. This is perhaps best represented in the work of Manuel Castells, and his contrasting of the network society to the pyramidal society of old (Castells, 2004). However, this narrative seems to be strongly contradicted by the fact that, similar to the ways in which many social movements evolve, political parties are in fact characterized by the opposite tendency, with the emergence of new forms of hierarchy and new forms of organization.

In the case of the digital party, this is exemplified by the phenomenon of hyper-leadership, namely the emergence of charismatic and personalized forms of leadership in new political organizations. In the case of digital parties this is seen in figures such as Podemos's leader Pablo Iglesias, France Insoumise's leader Jean-Luc Melenchon, and Five Star Movement's founder and guarantor Beppe Grillo and its leader Luigi di Maio. But it is a phenomenon that transcends digital parties, as is seen in a number of other movements. Take, for example, the case of Donald Trump and Alexandria Ocasio-Cortez in the United States, or Matteo Salvini in Italy, as social media-powered leaders who incarnate a highly personalized form of leadership. These and similar phenomena call for a redirection of current debates and for the acceptance that leadership, far from having disappeared, is resurfacing in new forms that mirror the specific characteristics of digital media.

However, also to be noted on this issue is that the open and unsolved questions are as many as those that have been resolved, at least in part. For a start we still do not have a clear picture of whether the emergence of new forms of charismatic and personalized leadership is an exceptional and transitory phenomenon or, rather, a trend that is here to stay. Moreover, it is not clear to what extent this phenomenon derives from technological factors or whether it is more of a consequence of the current phase of economic and political crisis, and of the populist trend that can be found across many current social and political phenomena.

How does leadership reflect new technological affordances? What are the different forms of leadership that have emerged in the present techno-political space? These are some of the questions that are still open to inquiry. Regarding the more general question of organizational structure and its topology, while it is apparent that the narrative of horizontality as a consequence of digital media has fallen flat, we lack a convincing and comprehensive alternative framework for understanding new organizational forms. What is the new dominant organizational format in the era of social media and platforms? Can notions of liquidity and flexibility explain what is at stake here? Or should we come up with a new language and new metaphors to capture the new organizational space? Can we, for example, opt for a concentric imaginary that, while allowing for horizontality and flexibility, reasserts the presence of hierarchies in the form of functional centres? These are some of the questions that many researchers are still grappling with, and it can be expected that it will take a long time to unlock some of them, as empirical phenomena related to them continue to evolve and, in so doing, throw up new research questions.

Conclusions

The return of political parties in the digital era calls for a highly perceptive analysis that may move us beyond the contemporary sense of surprise at this phenomenon and its novelty. We need to reconsider some of the key assumptions that were heretofore widely accepted in debates in political science and beyond.

New research in this area needs to grapple with the fact that many of the predictions made by the first wave of scholarship on digital politics were inaccurate, and that organizations continue to matter in the present era. This is particularly the case for political parties, which, far from having disappeared, are coming back, but coming back in very different forms from the previous era. Digital parties bear little resemblance to mass parties and their form of organization, and they therefore call for very different analytical grids to make sense of them and of their functioning.

In this chapter, I have highlighted some of the key issues that still remain open to scrutiny and which call for more systematic interventions. I have proposed that we need to look both at the bottom and the top of new forms of organization, which coincide with the study of new forms of participation and of leadership, respectively, as they are defined in the digital age. Future research will need to explore this question carefully, while researching the new political formations that are likely to arise in the coming years.

In developing this research endeavour, it is important to bear in mind two fundamental necessities: on the one hand, it is necessary to go beyond the anti-organizational suspicion that has so far been demonstrated to be misleading by the effective emergence of new political organizations, and to accept that traditional research problems, such as collectivity, power, leadership, and so on, remain as relevant as ever. On the other hand, it is advisable to be open-minded about the novelty of new digitally native organizations, bearing in mind that they will not necessarily be similar to the ones with which we are already familiar. Innovative research in digital politics will need to juggle the ability to engage with traditional sociological questions and the capacity to appreciate what is actually new about emerging digital formations. This mix of wisdom and open-mindedness will prove useful as we navigate the contradictory and rapidly transforming landscape of digital politics and the complex research questions it calls us to address.

References

Bennett, W.L. and Segerberg, A. (2012) 'The Logic of Connective Action: Digital Media and the Personalization of Contentious Politics', *Information, Communication and Society*, 15(5), pp. 739–68.

Castells, M. (2004) *The Network Society: A Cross-Cultural Perspective*. Cheltenham, UK and Northampton, MA, USA: Edward Elgar Publishing.

Chadwick, A. and Stromer-Galley, J. (2016) 'Digital Media, Power, and Democracy in Parties and Election Campaigns: Party Decline or Party Renewal?', *International Journal of Press/Politics*, 21(3), pp. 283–93.

Dalton, R.J. (2002) 'The Decline of Party Identifications'. In R.J. Dalton and M.P. Wateenberg (eds), *Parties without Partisans: Political Change in Advanced Industrial Democracies*. Oxford: Oxford University Press on Demand, pp. 19–36.

Deseriis, M. (2017) 'Direct Parliamentarianism: An Analysis of the Political Values Embedded in Rousseau, the "Operating System" of the Five STAR Movement'. *Proceedings of 2017 Conference for E-Democracy and Open Government (CeDEM)*, IEEE, May, pp. 15–25.

Gerbaudo, P. (2012) *Tweets and the Streets*. London: Pluto Press.

Gerbaudo, P. (2019) *The Digital Party: Political Organisation and Online Democracy*. London: Pluto Press.

Hartleb, F. (2013) 'Anti-Elitist Cyber Parties?', *Journal of Public Affairs*, 13(4), 355–69.

Helmond, A. (2015) 'The Platformization of the Web: Making Web Data Platform Ready', *Social Media+ Society*, 1(2). doi: 2056305115603080.

Karpf, D. (2012) *The MoveOn Effect: The Unexpected Transformation of American Political Advocacy*. New York: Oxford University Press.

Katz, R.S. and Mair, P. (1995) 'Changing Models of Party Organization and Party Democracy: The Emergence of the Cartel Party', *Party Politics*, 1(1), pp. 5–28.

Kirchheimer, O. (1966) 'The Transformation of the Western European Party Systems'. In J. LaPalombara and M. Weiner (eds), *Political Parties and Political Development*. Princeton, NJ: Princeton University Press, pp. 177–200.

Kreiss, D. (2015) 'The Problem of Citizens: E-Democracy for Actually Existing Democracy', *Social Media+ Society*, 1(2). doi: 2056305115616151.

Mair, P. (2013) *Ruling the Void: The Hollowing of Western Democracy*. London, UK and New York, USA: Verso Trade.

Margetts, H.Z. (2001) 'The Cyber Party: The Causes and Consequences of Organisational Innovation in European Political Parties', ECPR Joint Sessions of Workshops, Grenoble, 6–11 April.

Morozov, E. (2012) *The Net Delusion: The Dark Side of Internet Freedom*. New York: PublicAffairs.

Natale, S. and Ballatore, A. (2014) 'The Web Will Kill Them All: New Media, Digital Utopia, and Political Struggle in the Italian 5-Star Movement', *Media, Culture and Society*, 36(1), pp. 105–21.

Norris, P. (2001) *A Digital Divide: Civic Engagement, Information Poverty and the Internet in Democratic Societies*. New York: Cambridge University Press.

Panebianco, A. (1988) *Political Parties: Organization and Power*, Vol. 6. Cambridge: Cambridge University Press.

Shirky, C. (2008) *Here Comes Everybody: The Power of Organizing Without Organizations*. New York: Penguin.

Tufekci, Z. (2017) *Twitter and Tear Gas: The Power and Fragility of Networked Protest*. New Haven, CT: Yale University Press.

Van Biezen, I. and Piccio, D.R. (2013) 'Shaping Intra-Party Democracy: On the Legal Regulation of Internal Party Organizations'. In W. Cross and R.S. Katz (eds), *The Challenges of Intra-Party Democracy*. Oxford: Oxford University Press, pp. 27–48.

5 Researching the next wave of campaigns: empirical and methodological developments

Declan McDowell-Naylor

Introduction

The discussion in this chapter considers several empirical dimensions of digital campaigning that provide potential research focuses in the coming years, primarily involving political parties and their uses of digital media and technology. In addition, this chapter suggests some attendant methodological challenges and potential developments.

Empirical trends

Coleman and Freelon (2015: 4–7) have identified five general trends of 'an emerging digital political communication environment' which can serve to frame the discussion of campaigns. Firstly, overall control of the media agenda has become more difficult as sources of information and interaction multiply, meaning that a greater variety of communication techniques are required to reach citizens. Secondly, this more fluid system of communication is in tension with the hierarchical structures of existing political organizations, with unresolved consequences. Thirdly, as a result of these changes, co-existing notions of passive and active audiences exist, leading to further tensions in terms of communicative expectations. Fourthly, generic media rituals, such as broadcast interviews, have come to be seen by many audiences as stage-managed, leading to a greater emphasis on more idiomatic forms of communication. Fifthly, and finally, Coleman and Freelon argue that the digital environment appears 'invulnerable to strong legal regulation', in contrast with the age of mass media, but recognize it as an area of growing importance.

Campaigning is an important context in which the political tensions characterizing these trends manifest themselves, as digitally active actors seek to control the strategy and content of campaigns. Campaigns are also often the site of innovative forms of digital politics as parties and organizations seek new ways to reach citizens. Thus, current developments in campaigning across the world appear to suggest an intensification of these trends.

For example, in the two-year period between May 2017 and June 2019, various political parties from different nations (see Table 5.1) engaged in forms of digital campaigning and, generally, made crucial gains within their nation's elected legislature. These parties are politically distinct and culturally situated, yet they are all engaged in what Daniel Kreiss (2016: 3) has characterized as 'technology-intensive campaigning', in which 'parties and campaigns have invested considerable resources in technology, digital media, data, and analytics'. This is supported by an expanding body of academic research on how political parties and campaigns across the world have become increasingly digital (e.g., Baldwin-Philippi, 2015; Chadwick and Stromer-Galley, 2016; Dommett and Power, 2019; Gerbaudo, 2019; Gibson et al., 2017; Sampedro and Mosca, 2018). These studies have tended, however, to be overwhelmingly United States (US) and European focused. There is therefore an urgent need for research on developments in the Global South (Vaccari, 2019: 5), such as the digital innovations made by the BJP in India (Jaffrelot, 2015: 156–7).

Areas for further research

The first dimension to focus on is the role of consultancies. Consultancies received widespread attention during the 2018 Facebook–Cambridge Analytica data scandal. Despite this controversy, there remains a large commercial market available to those who are seeking to build digital campaigns. As Kreiss (2016: 194) has documented, consultancy firms offer a variety of tools and services, ranging from those that specialize in website building and email campaigns, to services such as polling, advertising and fundraising. The American firm Civis Analytics, for example, offers its clients data and behavioural analysis, predicative modelling, and marketing, all based on personal-level data (Civis Analytics, 2019). Moreover, the expertise of specific individuals, such as political adviser Jim Messina, is also being sought by campaigns. Indicating this trend, the early list of 2020 US Democratic candidates Cory Brooker,

Elizabeth Warren and Joe Biden were all reported to have paid Civis Analytics for services (Schwartz, 2019).

Table 5.1   Parties illustrative of digital forms of campaigning

Party (Country)	Election	Electoral +/-
Partido Social Liberal (Brazil)	2018 General Election	President: 55.13% (2nd round) Chamber: +44 Senate: +4
Labour Party (UK)	2017 General Election	Commons: +30
Brexit Party (UK)	2019 European Parliament	Parliament: +29
Bharatiya Janata Party (India)	2019 General Election	House: +21
Democratic Party (USA)	2018 Midterm Elections	House: +39 Senate: -2
Podemos (Spain)	2019 General Election	Congress: -29 Senate: -16
Movimento 5 Stelle (Italy)	2018 Italian General Election	Chamber: +114 Senate: +58

Many scholars point to the 2012 Barack Obama US presidential campaign as the point in which digital consultancy emerged (Cacciotto, 2017), although both Kreiss (2012) and Howard (2006) trace the emergence of the market throughout the 2000s. A complex network of US-based firms has since developed, meeting the demand for data-driven campaigning, but also raising concerns about data regulation. As this market evolves, research is needed to examine the interactions between consultancies and political campaigns, with social network analysis and elite interviews serving as potentially effective methods (see, e.g., Anstead, 2017; Lilleker et al., 2015).

In addition to outsourcing, many parties and organizations have also developed in-house digital tools and platforms. For example, during the 2019 Indian general election, the official Narendra Modi app (NaMo) provided users with news, updates and livestreams, as well as the ability to discuss, buy merchandise, donate and share preselected articles. In Italy, Movimento 5 Stelle's online platform, Rousseau, allowed supporters to vote on candidates and issues, while in the United Kingdom (UK), the pro-Labour grassroots organization

Momentum strategically directed activists during the 2017 general election using the digital tool My Nearest Marginal (Rees, 2017).

In response, Gerbaudo (2019: 4) has referred to the 'digital party' as a 'new organisational template', based upon digital technologies, that claims to be more democratic, more participatory and more transparent (Gerbaudo, Chapter 4 in this volume). NaMO, Rousseau and My Nearest Marginal are key examples of empirical case studies that can closely inform understanding of the trends identified by Freelon and Coleman. For example, while many of these tools and platforms allow for campaign activists to operate independently of the party hierarchy, they also retain certain affordances that allow political elites to guide behaviour, raising the question of power relations.

Data is the most fundamental aspect of digital campaigning. Data has always been central to political campaigning, since canvassing voters on doorsteps prior to elections and opinion polling goes back decades. It is the adoption of big data that means campaigns are becoming more sophisticated in the way that they collect and store individual-level data (Howard, 2006; Walker and Nowlin, 2018), leading to the analytics and micro-targeting increasingly characterizing election cycles. Data collection and maintenance is a costly and labour-intensive process, however, with parties needing to maintain up-to-date voter databases to ensure effective targeting during elections. Canvassing voters requires huge numbers of volunteers (Milne, 2019), while expensive digital advertisements can also facilitate data collection on responsive voters and be used to A/B test positive audience responses to thousands of message variations (Bedingfield, 2019). Individual-level data can also be purchased from various vendors, but again, at a cost.

Research is needed to both continue to understand the use of data in campaigns as methods develop, and examine how campaigns conceptualize citizens within their data (Kreiss, 2016: 209). Moreover, research is needed to explore the vast systems of data production, collection and sharing, and to begin analysing the political economy of digital campaigning (see Dommett and Power, 2019). In connection with mapping consultancy networks, this suggests a more critical turn in research approaches, and suggests that this new politics of data will require conceptual paradigms that go beyond communication (Koopman, 2019).

Online political advertising is being used extensively by campaigns across the world, leading to competition between them to control the media agenda (Thaker, 2019). For example, in the 2017 UK general election, the Conservative and Labour parties both paid for Google ads that appeared when searches for 'dementia tax' were made, in response to controversy that surrounded the policy (Booth and Mason, 2017). Online political advertising is being directly facilitated through technology firms (Kreiss and McGregor, 2018, 2019) such as Facebook and Google, whose own analytics enable micro-targeted political communication. According to Facebook's data,[1] political ad spending in the US dwarfs that of other countries, with spending in the UK, India and Spain being broadly comparable to one another. Many variations in message content are made possible by micro-targeted digital advertising, such as the 1433 variations revealed in BBC (2018) coverage of the official UK Vote Leave campaign, clearly showing how campaigns are adapting to the multiplications of information and interaction identified by Freelon and Coleman.

Research is needed to maintain reliable analyses of media effects and content as digital media outputs become ever more tailored to specific groups (see Hager, 2019; Borah, 2016). Moreover, research is needed to examine the consequences of the powerful commercial relationship that technology firms have formed with political campaigns (see Nielsen and Ganter, 2017). This is especially significant if online political ads become a focus of regulation (Kreiss and McGregor, 2019).

Social media has become an extensive aspect, not only of digital campaigning (Wells et al., 2016), but of the media system more generally (Chadwick, 2017). The prominence of certain political figures on social media is often the focus of research, as well as the use of social media by political parties (Magin et al., 2017; Stier et al., 2018) to get their messages to voters and citizens. In recent years, there has been a further diversification of social media, as political actors have been exploiting 'private' platforms such as WhatsApp and Instagram. WhatsApp has been used extensively to spread campaign messages – and misinformation – in both Brazil and India (Belli, 2018; Murgia et al., 2019), while in the United States, Instagram has been a prominent mode of communication, such as for the extremely popular Congresswoman Alexandria Ocasio-Cortez (Vo, 2019). Election candidates have been using social media as a way to explore new forms of communication ritual, as indicated by Coleman and Freelon, such as the use of quick, unsteady shots and emotional videos (D'Urso, 2019). Research is needed to examine these trends.

Despite correlated electoral gains and a strong commitment within political parties to digital campaigning, the effectiveness of data-driven campaigning techniques remains uncertain and understudied (Baldwin-Philippi, 2017; Endres and Kelly, 2018). Moreover, it is important to stress that digital campaigning is not a substitute for the traditional 'ground war' (Nielsen, 2012), and that campaign operatives continue to consider broadcast communication techniques as central to political campaigning (Hunter and Gladstone, 2019; Roose, 2019).

Nonetheless, successful campaign operatives have tended to publicly (and loudly) celebrate their digital methods as a strategic advantage. For example, during the 2019 European Parliament elections, Brexit Party leader Nigel Farage was quoted as saying, '[W]e have embraced modern social media perhaps in a way that no UK political party has ever done in the past. That's one reason that we are doing reasonably well' (Weston, 2019). Innovative digital campaigning methods thus fit easily within the horse-race framing of elections used by journalists, where effective data-driven vote targeting is portrayed as providing the candidate with a vital edge. It is, however, important that research does not conflate claims about effectiveness with empirical reality, but rather seeks to verify the effectiveness of these methods (see Hager, 2019).

The political consequences that evolving forms of digital campaigning entail, especially for democracy, are crucial to understand. This encompassing concern covers issues such as voters' freedom and rights online (Avila, 2019), secretive organizations engaging in 'stealth media' (Kim et al., 2018) and 'dark campaigning' (Moore, 2019), as well as more specific concerns, such as how Instagram's disappearing stories feature, where stories shared will disappear in a day, and Twitter's blocking feature which allows users to restrict what accounts can contact them, pose accountability issues within democratic society (Perry, 2018).

States are attempting to better grasp this situation. For example, in 2019, the UK's House of Lords Committee on Democracy and Digital Technologies announced an inquiry into 'the impact of digital technologies on political campaigning' (UK Parliament, 2019). Researchers should therefore expect to find digital politics research in increased demand in the coming years in relation to concerns about democratic society (see Chadwick et al., 2018; Gorton, 2016).

Finally, there is a question of power and control. The digitalization of campaigning has entailed the arrival of a new array of political actors. For example, skilled operatives are required to manage data within campaigns (Kreiss, 2016: 207), while digitally connected supporter networks have also emerged (Gibson et al., 2017). Both Momentum and the BJP are clear examples of how these supporter networks are used, and how hierarchical power structures within campaigns and parties are being transformed and challenged, as Freelon and Coleman identify. This leads to questions about a process of party renewal and adaption as digital participatory practices allow a greater array of actors into the party (Chadwick and Stromer-Galley, 2016). Similarly, alternative online political media organizations are also poised to challenge existing hierarchies (Holt et al., 2019), and this is another development that needs to be researched.

It is not possible to cover the entire range of potential research focuses. What is clear is that there are several fast-moving and intensifying trends, including populism and misinformation (which are not covered here), that require understanding through empirical findings. But there are important methodological challenges in researching the developments identified above.

Methodological issues

Firstly, a continued advancement of qualitative methods is required. Karpf et al. (2015: 1890) have argued for a 'new era of qualitative research', in which 'methods such as observation, participant observation, and in-field interviews, as well as in-depth interviews, focus groups, and process tracing' are used more consistently in order to keep pace with the trends discussed above. However, at this stage, qualitative research on campaigns has so far been concentrated within the US. What is needed is more qualitative research from within campaigns in locations such as the UK, as well as in the Global South, to emulate the kinds of research done in the US. The global nature of digital politics, with platforms such as WhatsApp available all over the world, gives this added importance.

Secondly, the ephemerality of political media broadcasts on platforms such as Instagram generates a need for different research techniques, such as 'live ethnography' (Chadwick, 2017), that can capture these flows of media in real time. Similarly, private messaging platforms entail many degrees of difficulty in terms of gaining access to and observing groups, as this depends on the ability of the researcher to cultivate an insider, participant–observer status. However, as digital campaigning expands infrastructural actors, it also expands the researcher's potential points of access.

Thirdly, cross-platform research is a vital area of development. While cross-platform research methods are both difficult and expensive to carry out, isolated studies of individual media platforms overlook the reality of a hybrid media system (Bode and Vraga, 2017) in which campaigns operate simultaneously across several platforms (see Bossetta, 2018, for an example). Advancing cross-platform methodology will require a careful understanding of the dynamic variety of affordances across platforms (Bode and Vraga, 2017: 2), as well as larger research teams working in a coordinated fashion.[2]

All methodologies are heavily dependent upon access to campaigns and the availability of data. The availability of access to Twitter, for example, has created an empirical bias within digital politics research (Karpf, 2016: 174) that will need to be overcome in order to study more private platforms such as Instagram or WhatsApp. Compounding this situation, research is entering a 'post-API age' of research (Freelon, 2018) as both Facebook and Twitter strictly limit and govern the availability of their application programming interfaces (APIs), thus making it increasingly difficult to get data formerly available to research digital politics and address democratic issues surrounding digital politics.

Notes

1. Following calls for greater transparency, Facebook has developed an advertisement library, which provides public access to advertisement data about social issues, elections or politics, including overall spending totals, spending by specific advertisers and spending data by geographic location.
2. Kreiss et al. (2018: 19) usefully define affordances as 'technological practices [that] are bounded by people's perceptions of what technologies can do, material or digital features that literally structure what can be done with them, and behaviours that emerge and evolve in relation to technologies' (see also Baym, 2010: 44).

References

Anstead, N. (2017) 'Data-Driven Campaigning in the 2015 United Kingdom General Election', *International Journal of Press/Politics*, 22(3), pp. 294–313.

Avila, R. (2019) 'Fixing Digital Democracy? The Future of Data-Driven Political Campaigning', *Open Democracy*, 7 January. Available at https://www.opendemocracy.net/en/fixing-digital-democracy-future-of-data-driven-political-campaigning/ (accessed 23 August 2019).

Baldwin-Philippi, J. (2015) *Using Technology, Building Democracy: Digital Campaigning and the Construction of Citizenship*. Oxford: Oxford University Press.

Baldwin-Philippi, J. (2017) 'The Myths of Data-Driven Campaigning', *Political Communication*, 34(4), pp. 627–33.

Baym, N.K. (2010) *Personal Connections in the Digital Age*. Cambridge: Polity Press.

BBC (2018) 'Targeted Pro-Brexit Facebook Ads Revealed', *BBC* News, 26 July. Available at https://www.bbc.com/news/uk-politics-44966969/ (accessed 3 September 2019).

Bedingfield, W. (2019) 'Boris Johnson's Facebook Advert Splurge Is All About Data', *Wired UK*, 5 August. Available at https://www.wired.co.uk/article/conservative-boris-johnson-facebook/ (accessed 6 August 2019).

Belli, L. (2018) 'Whatsapp Skewed Brazilian Election, Proving Social Media's Danger to Democracy', *Conversation*, 5 December. Available at http://theconversation.com/whatsapp-skewed-brazilian-election-proving-social-medias-danger-to-democracy-106476/ (accessed 2 September 2019).

Bode, L. and Vraga, E.K. (2017) 'Studying Politics Across Media', *Political Communication*, 35(1), pp. 1–7.

Booth, R. and Mason, R. (2017) 'Conservatives Buy "Dementia Tax" Google Ad as Criticism of Policy Grows', *Guardian*, 22 May. Available at https://www.theguardian.com/politics/2017/may/22/conservatives-buy-dementia-tax-google-ad-as-criticism-of-policy-grows/ (accessed 2 September 2019).

Borah, P. (2016) 'Political Facebook Use: Campaign Strategies Used in 2008 and 2012 Presidential Campaigns', *Journal of Information Technology and Politics*, 13(4), pp. 326–38.

Bossetta, M. (2018) 'The Digital Architectures of Social Media: Comparing Political Campaigning on Facebook, Twitter, Instagram, and Snapchat in the 2016 US Election', *Journalism and Mass Communication Quarterly*, 95(2), pp. 471–96.

Cacciotto, M.M. (2017) 'Is Political Consulting Going Digital?', *Journal of Political Marketing*, 16(1), pp. 50–69.

Chadwick, A. (2017) *The Hybrid Media System: Politics and Power*. Oxford: Oxford University Press.

Chadwick, A. and Stromer-Galley, J. (2016) 'Digital Media, Power, and Democracy in Parties and Election Campaigns: Party Decline or Party Renewal?', *International Journal of Press/Politics*, 21(3), pp. 283–93.

Chadwick, A., Vaccari, C. and O'Loughlin, B. (2018) 'Do Tabloids Poison the Well of Social Media? Explaining Democratically Dysfunctional News Sharing', *New Media and Society*, 20(11), pp. 4255–74.

Civis Analytics (2019) 'Products'. Available at https://www.civisanalytics.com/products/ (accessed 3 September 2019).

Coleman, S. and Freelon, D. (2015) 'Introduction'. In S. Coleman and D. Freelon (eds), *Handbook of Digital Politics*, Cheltenham, UK and Northampton, MA, USA: Edward Elgar Publishing, pp. 1–16.

Dommett, K. and Power, S. (2019) 'The Political Economy of Facebook Advertising: Election Spending, Regulation and Targeting Online', *Political Quarterly*, 90(2), pp. 257–65.

D'Urso, J. (2019) 'Tory Leadership: How Are Tory Hopefuls Campaigning Online?', *BBC News*, 2 June. Available at https://www.bbc.co.uk/news/uk-politics-48486562 (accessed 5 September 2019).

Endres, K. and Kelly, K.J. (2018) 'Does Microtargeting Matter? Campaign Contact Strategies and Young Voters', *Journal of Elections, Public Opinion and Parties*, 28(1), pp. 1–18.

Freelon, D. (2018) 'Computational Research in the Post-API Age', *Political Communication*, 35(4), pp. 665–8.

Gerbaudo, P. (2019) *The Digital Party*. London: Pluto Press.

Gibson, R., Greffet, F. and Cantijoch, M. (2017) 'Friend or Foe? Digital Technologies and the Changing Nature of Party Membership', *Political Communication*, 34(1), pp. 89–111.

Gorton, W.A. (2016) 'Manipulating Citizens: How Political Campaigns' Use of Behavioural Social Science Harms Democracy', *New Political Science*, 38(1), pp. 61–80.

Hager, A. (2019) 'Do Online Ads Influence Vote Choice?', *Political Communication*, 36(3), pp. 376–93.

Holt, K., Ustad Figenschou, T. and Frischlich, L. (2019) 'Key Dimensions of Alternative News Media', *Digital Journalism*, 7(7), pp. 1–10.

Howard, P.N. (2006) *New Media Campaigns and the Managed Citizen*. Cambridge: Cambridge University Press.

Hunter, F. and Gladstone, N. (2019) 'As Voters Tune Out, "Old Fashioned" Ad Campaigns Are Still in Favour', *Sydney Morning Herald*, 30 April. Available at https://www.smh.com.au/federal-election-2019/traditional-campaign-methods-still -in-favour-despite-parties-record-spending-on-digital-efforts-20190425-p51h5z .html/ (accessed 2 September 2019).

Jaffrelot, C. (2015) 'The Modi-Centric BJP 2014 Election Campaign: New Techniques and Old Tactics', *Contemporary South Asia*, 23(2), pp. 151–66.

Karpf, D. (2016) *Analytic Activism: Digital Listening and the New Political Strategy*. Oxford: Oxford University Press.

Karpf, D., Kreiss, D., Nielsen, R.K. and Powers, M. (2015) 'Qualitative Political Communication. Introduction – The Role of Qualitative Methods in Political Communication Research: Past, Present, and Future', *International Journal of Communication*, 9, p. 19.

Kim, Y.M., Hsu, J., Neiman, D., Kou, C., Bankston, L., Kim, S.Y., . . . Raskutti, G. (2018) 'The Stealth Media? Groups and Targets behind Divisive Issue Campaigns on Facebook', *Political Communication*, 35(4), pp. 515–41.

Koopman, C. (2019) 'Information before Information Theory: The Politics of Data beyond the Perspective of Communication', *New Media and Society*, 14(7), pp. 1164–180.

Kreiss, D. (2012) *Taking Our Country Back*. Oxford: Oxford University Press.

Kreiss, D. (2016) *Prototype Politics: Technology-Intensive Campaigning and the Data of Democracy*. Oxford: Oxford University Press.

Kreiss, D., Lawrence, R.G. and McGregor, S.C. (2018) 'In Their Own Words: Political Practitioner Accounts of Candidates, Audiences, Affordances, Genres, and Timing in Strategic Social Media Use', *Political Communication*, 35(1), pp. 8–31.

Kreiss, D. and McGregor, S.C. (2018) 'Technology Firms Shape Political Communication: The Work of Microsoft, Facebook, Twitter, and Google With Campaigns During the 2016 US Presidential Cycle', *Political Communication*, 35(2), pp. 155–77.

Kreiss, D. and McGregor, S.C. (2019) 'The "Arbiters of What Our Voters See": Facebook and Google's Struggle with Policy, Process, and Enforcement around Political Advertising', *Political Communication*, 36(4), pp. 1–24.

Lilleker, D.G., Tenscher, J. and Štětka, V. (2015) 'Towards Hypermedia Campaigning? Perceptions of New Media's Importance for Campaigning by Party Strategists in Comparative Perspective', *Information, Communication, and Society*, 18(7), pp. 747–65.

Magin, M., Podschuweit, N., Haßler, J. and Russmann, U. (2017) 'Campaigning in the Fourth Age of Political Communication: A Multi-Method Study on the Use of Facebook by German and Austrian Parties in the 2013 National Election Campaigns', *Information, Communication and Society*, 20(11), pp. 1698–1719.

Milne, O. (2019) 'How Labour beat Nigel Farage – Inside the Winning Peterborough Campaign', *Daily Mirror*, 7 June. Available at https://www.mirror.co.uk/news/politics/how-labour-beat-nigel-farage-16470450/ (accessed 3 September 2019).

Moore, M. (2019) 'Political Ads Are All over Facebook. But Voters Are in the Dark about Where They Come From', *The Guardian*, 4 August. Available at https://www.theguardian.com/commentisfree/2019/aug/04/facebook-ads-new-polticial-norm-voters-in-dark-where-they-come-from/ (accessed 6 August 2019).

Murgia, M., Findlay, S. and Schipani, A. (2019) 'India: The WhatsApp Election', *Financial Times*, 5 May. Available at https://www.ft.com/content/9fe88fba-6c0d-11e9-a9a5-351eeaef6d84/ (accessed 3 September 2019).

Nielsen, R.K. (2012) *Ground Wars: Personalized Communication in Political Campaigns*. Princeton, NJ: Princeton University Press.

Nielsen, R.K. and Ganter, S.A. (2017) 'Dealing with Digital Intermediaries: A Case Study of the Relations between Publishers and Platforms', *New Media and Society*, 20(4), pp. 1600–1617.

Perry, D.M. (2018) 'Alexandria Ocasio-Cortez Has Mastered the Politics of Digital Intimacy', *Pacific Standard*. Available at https://psmag.com/social-justice/alexandria-ocasio-cortez-has-mastered-the-politics-of-digital-intimacy/ (accessed 3 September 2019).

Rees, E. (2017) 'What Made the Difference for Labour? Ordinary People Knocking on Doors', *Guardian*, 12 June. Available at https://www.theguardian.com/commentisfree/2017/jun/12/labour-knocking-on-doors-jeremy-corbyn-momentum (accessed 3 September 2019).

Roose, K. (2019) 'Digital Hype Aside, Report Says Political Campaigns Are Mostly Analog Affairs', *New York* Times, 21 March. Available at https://www.nytimes.com/2019/03/21/business/political-campaigns-digital-advertising.html/ (accessed 3 September 2019).

Sampedro, V. and Mosca, L. (2018) 'Digital Media, Contentious Politics and Party Systems in Italy and Spain', *Javnost – The Public*, 25(1–2), pp. 160–68.

Schwartz, B. (2019) 'Joe Biden's Campaign Taps Data Analytics Company Backed by Former Google Chairman Eric Schmidt', *CNBC*, 29 April. Available at https://www.cnbc.com/2019/04/29/biden-taps-data-analytics-company-backed-by-eric-schmidt-for-digital-resources.html/ (accessed 3 September 2019).

Stier, S., Bleier, A., Lietz, H. and Strohmaier, M. (2018) 'Election Campaigning on Social Media: Politicians, Audiences, and the Mediation of Political Communication on Facebook and Twitter', *Political Communication*, 35(1), pp. 50–74.

Thaker, A. (2019) 'Modi's BJP Is Back to Being the Top Facebook Advertiser amid Indian Elections', *Quartz India*, 17 April. Available at https://qz.com/india/1591629/modis-bjp-tops-congress-in-political-ads-on-facebook-in-india/ (accessed 3 September 2019).

UK Parliament (2019) 'Should Democracies Embrace or Fear Digital Technologies? New Lords Committee Seeks Your Views'. Available at https://www.parliament.uk/business/committees/committees-a-z/lords-select/democ-digital-committee/news-parliament-2017/call-for-evidence-digital/.

Vaccari, C. (2019) Editorial. *International Journal of Press/Politics*, 24(1), pp. 3–6.

Vo, L.T. (2019) 'Why Is Everyone So Obsessed With AOC? Let's Analyze The Memes', *Buzzfeed News*. Available at https://www.buzzfeednews.com/article/lamvo/alexandria-ocasio-cortez-aoc-conservatives-liberals-meme/ (accessed 3 September 2019).

Walker, D. and Nowlin, E.L. (2018) 'Data-Driven Precision and Selectiveness in Political Campaign Fundraising', *Journal of Political Marketing*, pp. 1–20. 10.1080/15377857.2018.1457590.

Wells, C., Shah, D.V., Pevehouse, J.C., Yang, J., Pelled, A., Boehm, F., . . . Schmidt, J.L. (2016) 'How Trump Drove Coverage to the Nomination: Hybrid Media Campaigning', *Political Communciation*, 33(4), pp. 669–76.

Weston, K. (2019) 'Brexit News: Nigel Farage Reveals KEY Tactic that Helped Brexit Party', *Daily Express*, 28 April. Available at https://www.express.co.uk/news/uk/ 1120043/brexit-news-latest-party-nigel-farage-iain-dale-change-uk-eu-elections -poll-live-today-uk/ (accessed 3 September 2019).

6 Digital advertising in political campaigns and elections

Laleah Fernandez

Introduction

The term 'psychographic profiling' became the topic of headlines to describe tactics behind Cambridge Analytica's role in the 2016 United States (US) presidential election. The scandal, along with the subsequent election of Donald Trump, brought to light the emerging significance of data analytics, political advertising and the exploitation of social media users' data in political campaigns and elections.

Many other examples have occurred before and since the Cambridge Analytica case. Barack Obama's campaigns gained notoriety for combining voter records and social media data to reach key groups and craft messaging in the 2008 and 2012 US presidential elections (Wortham, 2012). In the 2015 United Kingdom general election campaign, the Conservative Party purchased Facebook users' geographical data combined with private tracker polls in key swing constituencies to identify and target undecided voters with specific concerns and behavioural traits (Chadwick and Stromer-Galley, 2016). Jean-Luc Mélenchon and his movement, La France Insoumise, used online mobilization and street protests similar to those used by Italy's Five Star Movement, and the aggressive Internet campaigns of the far-right German party AfD. The use of digital political advertising is a growing approach to targeting that is perceived to be effective by politicians and campaigns around the world.

While advertisers generally target consumers as groups, political campaigns need to target specific people, registered voters receptive to a potential message. As such, online advertisers have strongly resisted any measures that would allow users to avoid tracking (Hoofnagle et al., 2012: 275). The appeal of Cambridge Analytica's psychographic profiling was in the promise to combine the precision of data analytics with the insights of behavioural psychology and the best of individually addressable advertising technology. The company's website promised tools and data to 'run a truly end-to-end campaign'.

The increased reliance on digital media for political advertising raises the importance of identifying and addressing gaps in academic research. These include research on the effectiveness of digital political messaging and persuasion tactics; identifying challenges to traditional campaigns and brand management strategies; and researching the effects of trolls, fake and bot accounts on political discourse and political engagement, as well as the longer-term viability of digital advertising. In addition, research intended to inform regulation and public policy options is largely absent from the existing body of academic scholarship.

The role of digital media in shaping innovations in political advertising

The increasing use of technology in politics has had a profound effect on political advertising in the last decade (Van Steenburg, 2015). Traditionally, campaigns relied on a set of standard distribution channels ranging from shaking hands and speeches at rallies, to billboards, radio and TV ads. Historically, media served a gatekeeping role and costs associated with traditional forms of political advertising served as a major barrier for lesser-known candidates and interest groups. Media gatekeepers and cost barriers resulted in a few candidates receiving a disproportionate share of attention on television and radio.

The Internet lowers the barriers of entry, allowing even minor candidates without institutional support to broadcast messages quickly and inexpensively (Wattal et al., 2010). Campaigns now use digital tools to influence media coverage and public perceptions, or to bypass professional news media altogether by speaking directly to their supporters (Jungherr, 2016). Sharing practices of socially networked citizens is another route for getting around media gatekeepers. Political content can be 'liked', commented on or shared, making the message more relevant than traditional political advertisements. When content is shared within the target market, it adds higher credibility than that offered by candidate or campaign. Digital advertising allows for a broader reach with targeted campaign appeals aimed at improving voter engagement, fundraising and generating voter turnout (Dutton, 2010), disrupting the tried and true approaches of persuasive political communication historically dominated by political elites and their operatives (Dutton, 2010; Penney, 2017).

Online technologies also empower the role of the average networked citizen (Dubois and Dutton, 2014; Dutton, 2015). Potential voters have a multitude of new public spaces for political engagement, ranging from blogging and social networks to online polls or live commentary on debates. Unlike traditional forms of political advertising, social media facilitates contact with politicians, and instant feedback (Dutton, 2010), enabling supporters to interact directly and easily with campaigns. All of this creates a widely visible and immediate way for potential voters to respond to candidate messages and publicly express support. These are only a handful of the many ways in which the Internet facilitates the exchange of viewpoints and can bring disengaged citizens into political life (Dubois and Dutton, 2014).

Beyond cost of entry and resource limitations, traditional campaigns have a set of other limitations as a result of the gatekeeping roles of legacy news outlets. More specifically, news outlets could choose what to report, what to omit and the tone of the message. Arguably, traditional legacy media's more balanced coverage of political issues limited the extent and extremity of political speech to prompt mobilization or participation (Bruns, 2011). Traditional, mainstream news outlets publish less polarizing and opinionated content than some alternative digital information sources (Ohme, 2019), as such content from blogs or political social media pages tends to be more partisan and more focused on mobilizing potential voters (Stroud, 2010).

Building a research agenda

Scholars need a research agenda that integrates a concern for the changing practices of online political advertising with the direct, indirect, intentional and unintentional impacts on individuals and democracy. Campaigns are not only spending more on digital advertising (Williams and Gulati, 2018), but they are also increasingly more creative in their media mix and use of analytics to reach and mobilize key groups. These types of new media tools are generally seen as playing a complementary role to traditional media, in what Chadwick (2013) has termed the hybrid media system, characterized by interaction between older and newer media logics.

Beyond the move to a more hybrid mix of traditional and digital media in political advertising, the use of analytics by campaigns marks a pivotal turn in the nature of elections worldwide. Analytics are used to aggregate behavioural information from public voter records and digital media environments, with the aim of organizing and mobilizing key segments of the electorate to vote and to share their decision with others (Chadwick and Stromer-Galley, 2016). This has led to innovative targeting techniques that include, but are not limited to, micro-targeting, geolocation-targeting and targeting across devices.

Micro-targeting spans devices and mediums, for example a US platform owned by Comcast, Cox and Spectrum provides campaigns with the ability to target potential voters through viewing data combined with demographic and cross-platform data (Chester and Montgomery, 2017). Geolocation targeting is possible as mobile devices continually send location tracking signals through Wi-Fi, Bluetooth and the Global Positioning System (GPS). By matching email addresses with cookies, Internet Protocol (IP) addresses, and other identifiers, cross-device recognition is a technique in which campaign operatives link an individual across social networks, to a personal computer and/or mobile device (Díaz-Morales, 2015; Levine, 2016). Campaigns can find targets on mobile devices at specific times when mobile phone users may be more receptive to a message, at the same time using voter data to enable tailored advertisements (Kaye, 2016).

Targeting services are readily available to optimize these techniques with the help of specialty firms. With an increasing number of cloud-based tools developed by companies such as Adobe, Oracle, Salesforce, Nielsen and IBM, the practice of selling political data and detailed consumer information is now commonplace for political advertising. Such user data ranges from credit card use, personal interests, consumption patterns and TV viewing patterns (Chester and Montgomery, 2017). In some instances, campaigns have also been known to use these practices to demobilize or suppress targeted voters, such as by crafting personalized messages indicating long queues at local ballot stations (Green and Issenberg, 2016).

Empirical and scholarly research is indeed straddled by a number of key challenges that in the era of digital politicking are complicated. Beyond intentional voter suppression, other means of pushing out persuasive content such as political bots, or automated scripts run by political actors, are increasingly used to manipulate public opinion over social media (Nee and De Maio, 2019), not to mention the increased use of anonymous groups and 'dark money' campaigns, warranting further investigation to better understand the significance of digital political advertising in contemporary democracy, and bringing issues such as transparency and accountability into the broader conversation (Kim et al., 2018).

Research agenda I: marketing mix and advertising effectiveness

Digital political advertising studies often fall into two broad categories of supply-side and demand-side studies (Xenos et al., 2017). Demand-side research focuses more heavily on potential impacts and forms of participation (Sudulich and Wall, 2010). On the demand side of political advertising, few studies have investigated the relationship between the use of different forms of digital media and their relative effects on political participation and knowledge (Dimitrova et al., 2014). Nor has research in this area established a clear differentiation between the effects of political marketing and the other factors that influence voters (Van Steenburg, 2015). Further, demand-side studies arguably provide a limited picture of the dynamics of candidate and supporter interaction and engagement (Xenos et al., 2017). Further, research in this area has not established clear differentiation between the effects of political marketing and the other factors that influence voters (Van Steenburg, 2015).

Supply-side studies examine new digital tools and how are they being used. Such research looks at which candidates, under which circumstances, use particular online campaigning features (Gibson, 2012). On the supply side, research would benefit from the expansion of marketing and persuasion models to better incorporate the changing dynamics of the digital sphere. In general, the political advertising literature has not sufficiently applied consumer behaviour insights to voting behaviour (Van Steenburg, 2015). Overall, there remains a need for a better understanding of the cognitive effects of digital media (Dimitrova et al., 2014) and the broadening of conceptual and empirical models of electoral outcomes and campaign effects (Gibson, 2012). Theories such as social learning theory (Bandura, 1977), the theory of reasoned action (Fishbein and Ajzen, 2010) and cognitive dissonance (Festinger, 1962/1985 [2001]) are all examples of cognitive decision models that can provide a framework for formative research, strategy development and campaign evaluation. Future studies would be well served to leverage and modify such models, while developing new theories to better fit an increasingly digitized political environment.

Applied research in this area could build on such theories to offer specific techniques and approaches to broaden reach and engagement, while maintaining control of a political brand and messaging. For example, social media platforms play a significant role in the media diet of digital native voters and can foster campaign participation (Ohme, 2019), yet potential voters are no longer passive participants, creating a tension between leveraging interactivity and the control of content (Freelon, 2017). The hybrid nature of political advertising, in combination with the pace of technology change and availability of online tools, calls for mixed methodological approaches to better understand trends and changes over time. Thus, scholars need to consider not only quantitative measures available through trace data and Web-scraping, but also ethnographic fieldwork (Laaksonen et al., 2017) and qualitative approaches such as interviews and surveys of practitioners and potential voters. Surveys can be employed and split into exposure to different media types, but should not be isolated to a single media, platform or type of exposure, to offer more holistic findings that encompass a mix of mediums.

While strategy and branding are of utmost importance in the development and execution of political campaigns, the effectiveness of candidate brands remains unclear (Van Steenburg, 2015). Such gaps can be overcome by a closer examination of voters as consumers. However, there are a number of key challenges that must be overcome to study brand management efficiently in the digital political sphere. For instance, online tools allow for the participation in political discourse with a higher degree of anonymity than in the past. With anonymity comes confidence to act and behave in ways that are at times ethically questionable, illegal or exploitative (Hughes, 2018). Such content can quickly go viral amid the instantaneous and continuous cycle of dissemination and consumption of political content. Ways in which candidates or campaigns can effectively respond to extreme content or correct misinformation or disinformation is an area of research that deserves more scholarly attention.

Research agenda II: data ethics and targeted manipulation

Van Dijck and Poell (2015) argue that connected citizens have entered a platform society, characterized by a global conglomerate of interdependent platforms that are transforming the political economy of the media landscape (van Dijck and Poell, 2015: 1). With this entanglement of platforms, researchers too often overlook the role and consequences of seemingly innocuous design decisions made at the platform level. For example, social media platforms supply communication resources that allow bots to escape detection and enact influence. It is estimated that leading up to the 2016 US presidential election, Twitter was populated by nearly 50 million bot accounts (Varol et al., 2017). Given the gatekeeping power of technology firms to either inhibit or enhance what people are exposed to, it is essential to develop a more thorough understanding of company policies that either restrict or amplify the flow of certain types of information. Comparative research looking at the role of technology firms and their agenda is needed to fill this gap.

Future research should examine how data are collected and interpreted by political campaigns to learn about voters' political preferences and to inform campaign strategies and priorities, including creating voter profiles and testing campaign messaging. Controversies over the 2016 election have triggered greater public scrutiny over some of the practices that have become standard operating procedures in the digital media and marketing ecosystem. These procedures include micro-targeting, cross-device targeting and geolocation targeting. The extent to which stores of existing data on potential voters is exchanged between political candidates, acquired from national repositories, or sold to those who want to leverage them, remains unclear. The value of these data stores is indisputable, and campaigns are highly motivated to gather such political intelligence to gain an edge. In this vein, we are likely to see more research examining the ethical implications of how data are collected, analysed and used to target and reach potential voters with the aim of influencing or manipulating their views or voting behaviour.

For academics, as well as political operatives, data storage and retrieval capabilities make information available that was once hidden. Social media provide a robust repository of data where messages are recorded, and in some cases more easily accessed, making systematic quantitative and qualitative content analysis of communication content more feasible, albeit limited by issues of personal privacy, access to data and platform-level policies. Access to digital trace data depends on the provisions of corporations holding the data; enterprises whose policies on data storage, retention and access provision for researchers must follow commercial, operational, ethical and legal considerations (Jungherr, 2016). For example, there is currently no way to independently extract content from Facebook without violating its terms of service, and companies can restrict or eliminate application programming interface (API) access at any time (Freelon, 2018). As Jungherr points out, the age of big data requires that researchers streamline efforts to develop standards for data collection, preparation, analysis and reporting.

Research agenda III: platform policies, regulations and public policy

Research has an important role to play in the policy discussion, especially with new and emerging media. Scholars also have an opportunity to study the impact of trolls, bots and algorithms, and to help clarify the respective roles and responsibilities of platform operators. Such social bots are accounts run by automated software that mimic real users, used to automate group pages and spread political advertisements. Bots become agents by harnessing profile settings, popularity measures and automated conversation tools, along with vast amounts of user data that social media platforms make available. Extreme content is also amplified as trolls, bots and algorithms exploit measurement metrics of digital media platforms (Unver, 2016). Yet, studies on whether policies have been implemented, or on their effects, are largely lacking in the existing body of research. Research coordinated with advocates and policy-makers can facilitate a more central role in the public policy debate, helping to define the issues, identify priorities and provide a framework for evaluating policy alternatives.

Firstly, if scholars are to have an effect on policy, research must closely monitor and measure technological changes and their impact on individuals and democracy. This extends to all levels of public policy as well as platform policies. Future research should further examine the role of company policies with an eye on possible contradictions between law, regulation and self-regulation (Noto La Diega, 2018). Secondly, longitudinal studies of the impact of political advertising are important, as well as policy-relevant research that addresses specific issues and needs. At the same time, building from theory-based research, such as cognitive decision models, is another way to appeal to policy-makers by finding mechanisms to address individual behaviour change rather than relying solely on regulation.

Lastly, researchers can play a role in public education, and pressure governments to design and launch educational campaigns to help users understand the value of their data, the meaning and consequences of digital political advertising in general, and the mechanisms behind micro-targeting efforts in particular. For example, in the US, digital ads require no disclosure of the paying party, making the goal or political agenda of ads less discernible. At the same time, many questions remain about what constitutes an ad. These questions are complicated in an era ripe for political micro-targeting, a technique in which user data are harvested to create voter profiles. In order to make informed decisions, voters need information about who pays for online political ads, and what content is designed for persuasive rather than informational purposes.

Continued research is needed to help policy-makers understand how consumer privacy interventions can enable choice, and the extent to which marketplace approaches can deny consumers opportunities to exercise autonomy

(Hoofnagle et al., 2012: 274). For example, direct political communication can be established if voters follow political candidates or parties on social media as well as through micro-targeting, which increasingly takes place on these platforms (Bimber, 2014). This practice, when used for telemarketing and direct mail, was met with a tremendous public backlash, leading to a series of regulations and tools intended to protect consumers from unsolicited intrusions. To enable choice, consumers needed legal rules which allowed for opting out of telemarketing.

Academics must also take a stand in the policy realm to address the practical and methodological implications of digital advertising practices. The current state of political advertising practices poses significant methodological challenges to research. Most notably, the effects of dark money, trolls, fake accounts and bots are difficult to identify for analysis. Researchers have pointed to the need for a broader exploration of the roles and impact of bot accounts in academic research, and the possibility of inflated or inaccurate findings that result (Larsson, 2019). This call for broader research should be extended to clear directives for best practices, not only for the public good but also toward the accuracy and feasibility of academic research. Interwoven with these considerations, scholars must continue to identify what policies are most important to the development of a healthy, democratic media system, and how to mitigate, reduce or avoid harm. As regulatory agencies take up policy issues, researchers must be prepared to respond with data, solutions and approaches that can inform decision-making.

References

Bandura, A. (1977) 'Self-Efficacy: Toward a Unifying Theory of Behavioral Change', *Psychological Review*, 84(2), pp. 191–215. Available at https://psycnet.apa.org/record/1977-25733-001/ (accessed 8 November 2019).

Bimber, B. (2014) 'Digital Media in the Obama Campaigns of 2008 and 2012: Adaptation to the Personalized Political Communication Environment', *Journal of Information Technology and Politics*, 11, pp. 130–50. Available at https://doi.org/10.1080/19331681.2014.895691/ (accessed 8 November 2019).

Bruns, A. (2011) 'Gatekeeping, Gatewatching, Real-Time Feedback: New Challenges for Journalism', *Brazilian Journalism Research*, 7, pp. 117–36. Available at https://doi.org/10.25200/BJR.v7n2.2011.355/ (accessed 8 November 2019).

Chadwick, A. (2013) *The Hybrid Media System: Politics and Power*, Oxford Studies in Digital Politics. Oxford, UK and New York, USA: Oxford University Press.

Chadwick, A. and Stromer-Galley, J. (2016) 'Digital Media, Power, and Democracy in Parties and Election Campaigns: Party Decline or Party Renewal?', *International*

Journal of Press/Politics, 21, pp. 283–93. Available at https://doi.org/10.1177/1940161216646731/ (accessed 8 November 2019).

Chester, J. and Montgomery, K.C. (2017) 'The Role of Digital Marketing in Political Campaigns', *Internet Policy Review*, 6(4). Special Issue.

Díaz-Morales, R. (2015) 'Cross-Device Tracking: Matching Devices and Cookies', *2015 IEEE International Conference on Data Mining Workshop (ICDMW)*, pp. 1699–704. Available at https://doi.org/10.1109/ICDMW.2015.244/ (accessed 8 November 2019).

Dimitrova, D.V., Shehata, A., Strömbäck, J. and Nord, L.W. (2014) 'The Effects of Digital Media on Political Knowledge and Participation in Election Campaigns: Evidence From Panel Data', *Communication Research*, 41, pp. 95–118. Available at https://doi.org/10.1177/0093650211426004/ (accessed 8 November 2019).

Dubois, E. and Dutton, W.H. (2014) 'Empowering Citizens of the Internet Age'. In M. Graham and W.H. Dutton (eds), *Society and the Internet*. Oxford: Oxford University Press, pp. 238–54. https://doi.org/10.1093/acprof:oso/9780199661992.003.0016/ (accessed 8 November 2019).

Dutton, W.H. (2010) 'Democratic Potential of the Fifth Estate', *PerAda Magazine*. Available at http://citeseerx.ist.psu.edu/viewdoc/download?doi=10.1.1.654.1371&rep=rep1&type=pdf/ (accessed 8 November 2019).

Dutton, W.H. (2015) 'The Internet's Gift to Democratic Governance: The Fifth Estate'. In S. Coleman, G. Moss and K. Parry (eds), *Can the Media Serve Democracy?* London: Palgrave Macmillan, pp. 164–73.

Festinger, L. (1962/1985 [2001]) *A Theory of Cognitive Dissonance*. Reissued by Stanford University Press in 1962, renewed 1985 by author (reprint). Stanford, CA: Stanford University Press.

Fishbein, M. and Ajzen, I. (2010) *Predicting and Changing Behavior: The Reasoned Action Approach*. New York: Psychology Press.

Freelon, D. (2017) 'Campaigns in Control: Analyzing Controlled Interactivity and Message Discipline on Facebook', *Journal of Information Technology and Politics*, 14(2), pp. 168–181. https://doi.org/10.1080/19331681.2017.1309309/ (accessed 8 November 2019).

Freelon, D. (2018) 'Computational Research in the Post-API Age', *Political Communication*, 35, pp. 665–8. Available at https://osf.io/preprints/socarxiv/56f4q/ (accessed 8 November 2019).

Gibson, R. (2012) 'From Brochureware to "MyBo": An Overview of Online Elections and Campaigning', *Politics*, 32, pp. 77–84. Available at https://onlinelibrary.wiley.com/doi/abs/10.1111/j.1467-9256.2012.01429.x/ (accessed 8 November 2019).

Green, J. and Issenberg, S. (2016) 'Why the Trump Machine Is Built to Last Beyond the Election', *Bloomberg.Com*, 27 October. https://www.bloomberg.com/news/articles/2016-10-27/inside-the-trump-bunker-with-12-days-to-go.

Hoofnagle, C.J., Soltani, A., Good, N. and Wambach, D.J. (2012) 'Behavioral Advertising: The Offer You Can't Refuse', *Harvard Law Policy Review*, 6, pp. 273–96.

Hughes, A. (2018) *Market Driven Political Advertising: Social, Digital And Mobile Marketing*, Palgrave Studies in Political Marketing and Management. Cham: Palgrave Macmillan.

Jungherr, A. (2016) 'Four Functions of Digital Tools in Election Campaigns: The German Case', *International Journal of Press/Politics*, 21, pp. 358–77. Available at https://www.researchgate.net/publication/300082509_Four_Functions_of_Digital_Tools_in_Election_Campaigns_The_German_Case/ (accessed 8 November 2019).

Kaye, K. (2016) 'RNC's Voter Data Provider Teams Up With Google, Facebook and Other Ad Firms', *AdAge*, 15 April. Available at http://adage.com/article/campaign -trail/rnc-voter-data-provider-joins-ad-firms-including-facebook/303534/ (accessed 8 November 2019).

Kim, Y.M., Hsu, J., Neiman, D., Kou, C., Bankston, L., Kim, S.Y., . . . Raskutti, G. (2018) 'The Stealth Media? Groups and Targets behind Divisive Issue Campaigns on Facebook', *Political Communication*, 35, pp. 515–41. Available at https://doi.org/10 .1080/10584609.2018.1476425/ (accessed 8 November 2019).

Laaksonen, S.-M., Nelimarkka, M., Tuokko, M., Marttila, M., Kekkonen, A. and Villi, M. (2017) 'Working the Fields of Big Data: Using Big-Data-Augmented Online Ethnography To Study Candidate–Candidate Interaction at Election Time', *Journal of Information Technology and Politics*, 14, pp. 110–131. Available at https://www .tandfonline.com/doi/abs/10.1080/19331681.2016.1266981?journalCode=witp20/ (accessed 8 November 2019).

Larsson, A.O. (2019) 'Winning and Losing on Social Media: Comparing Viral Political Posts across Platforms', *Convergence: The International Journal of Research into New Media Technologies*. Available at https://www.researchgate.net/publication/ 329196134_Winning_and_Losing_on_Social_Media_-_Comparing_Viral_Political _Posts_Across_Platforms/ (accessed 8 November 2019).

Levine, B. (2016) 'Report: What is Data Onboarding, and Why Is It Important to Marketers?', *MarTech Today*. Available at https://martechtoday.com/report-data -onboarding-important-marketers-192924/ (accessed 8 November 2019).

Nee, R.C. and De Maio, M. (2019) 'A "Presidential Look"? An Analysis of Gender Framing in 2016 Persuasive Memes of Hillary Clinton', *Journal of Broadcasting and Electronic Media*, 63, pp. 304–21. Available at https://doi.org/10.1080/08838151 .2019.1620561/ (accessed 8 November 2019).

Noto La Diega, G. (2018) 'Some Considerations on Intelligent Online Behavioural Advertising', *Revue du droit des technologies de l'information*, 66–7, pp. 53–90.

Ohme, J. (2019) 'When Digital Natives Enter the Electorate: Political Social Media Use among First-Time Voters and Its Effects on Campaign Participation', *Journal of Information Technology and Politics*, 16(2), pp. 119–36. Available at https://doi.org/ 10.1080/19331681.2019.1613279/ (accessed 8 November 2019).

Penney, J. (2017) *The Citizen Marketer: Promoting Political Opinion in the Social Media Age*, Oxford Studies in Digital Politics. New York: Oxford University Press.

Stroud, N.J. (2010) 'Polarization and Partisan Selective Exposure', *Journal of Communication*, 60, pp. 556–76. Available at https://doi.org/10.1111/j.1460-2466 .2010.01497.x/ (accessed 8 November 2019).

Sudulich, M.L. and Wall, M. (2010) '"Every Little Helps": Cyber-Campaigning in the 2007 Irish General Election', *Journal of Information Technology and Politics*, 7, pp. 340–55. Available at https://doi.org/10.1080/19331680903473485/ (accessed 8 November 2019).

Unver, H.A. (2016) 'Digital Challenges to Democracy: Politics of Automation, Attention, and Engagement', *Journal of International Affairs*, 71(1), pp. 127–46.

van Dijck, J. and Poell, T. (2015) 'Social Media and the Transformation of Public Space', *Social Media + Society* 1(2). Available at https://journals.sagepub.com/doi/full/10 .1177/2056305115622482/ (accessed 8 December 2019).

Van Steenburg, E. (2015) 'Areas of Research in Political Advertising: A Review and Research Agenda', *International Journal of Advertising*, 34, pp. 195–231. Available at https://www.tandfonline.com/doi/abs/10.1080/02650487.2014.996194/ (accessed 8 December 2019).

Varol, O., Ferrara, E., Davis, C.A., Menczer, F. and Flammini, A. (2017) *Online Human-Bot Interactions: Detection, Estimation, and Characterization.* arXiv:1703 .03107 [cs]. http://arxiv.org/abs/1703.03107.

Wattal, S., Schuff, D., Mandviwalla, M. and Williams, C.B. (2010) 'Web 2.0 and Politics: The 2008 US Presidential Election and an E-Politics Research Agenda', *MIS Quarterly*, 34, p. 669. Available at https://doi.org/10.2307/25750700/_(accessed 8 December 2019).

Williams, C.B. and Gulati, G.J. (2018) 'Digital Advertising Expenditures in the 2016 Presidential Election', *Social Science Computer Review*, 36, pp. 406–21. Available at https://journals.sagepub.com/doi/abs/10.1177/0894439317726751?journalCode= ssce/ (accessed 8 December 2019).

Wortham, J. (2012) 'Campaigns Use Social Media to Lure In Younger Voters', *New York Times*, 7 October. https://www.nytimes.com/2012/10/08/technology/campaigns-use -social-media-to-lure-younger-voters.html.

Xenos, M.A., Macafee, T. and Pole, A. (2017) 'Understanding Variations in User Response to Social Media Campaigns: A Study of Facebook Posts in the 2010 US Elections', *New Media and Society*, 19, pp. 826–42. Available at https://journals .sagepub.com/doi/abs/10.1177/1461444815616617?journalCode=nmsa/ (accessed 8 December 2019).

7 The role of digital media in China: participation in an unlikely place

Wan-Ying Lin and Xinzhi Zhang

Introduction

This chapter reviews studies on how individuals participate in different types of political engagement through digital media in mainland China, where channels for conventional political engagement are restricted and Internet censorship and surveillance are prevailing practices (Freedom House, 2019). This chapter discusses three questions:

1. What are the major types of political engagement in mainland China?
2. What is the role of digital media, especially online media, such as the Internet and social networking sites (SNS), in the ways people engage in politics, and how are these relationships similar to or different from other countries and societies?
3. On this basis, given the rapidly changing media and political landscape in mainland China, what are some of the key research questions on digital politics?

Types of political engagement in mainland China

Political engagement refers to all kinds of citizens' actions 'to communicate information to government officials about their concerns and preferences and to put pressure on them to respond' (Verba et al., 1995: 37). In mainland China, the ruling authorities have ardently defended the status quo and have opposed conventional political actions under a single-party political system (Kuan and Lau, 2002). Therefore, building on previous studies, we classify political engagement based on two distinct dimensions: (1) the relationship between the government and citizens, whether the actions are elite-challenging or elite-directed; and (2) how the engagement is manifested, whether it is explicit or implicit (see Table 7.1).

Table 7.1   A typology of political engagement

| | | Government–citizen relationship | |
		Elite-challenging	Elite-directed
Level of manifestation	Explicit	Public resistance against the ruling authorities (e.g., large-scale demonstrations and protests)	Political engagement organized by the ruling authorities (e.g., elections, official celebrations)
	Implicit	Resistance without openly challenging the status quo (e.g., online parody)	Campaigns organized by the ruling authorities aiming at reinforcing their legitimacy (e.g., environmental campaigns organized by government-led non-governmental organizations)

For the government–citizen relationship, Inglehart (1977: 299) termed institutional political engagement as 'elite-directed political participation' when it is mobilized by 'typically hierarchical organizations in which a small number of leaders (lead) a mass of disciplined troops'. These political actions are organized by and proceed under the channels and scopes predetermined by the ruling authorities. Unconventional and non-institutionalized political engagement, on the other hand, is defined as 'elite-challenging political participation', which is based less on established bureaucratic organizations and more on ad hoc groups that aim to effect specific policy change (Inglehart, 1977: 300).

With respect to how the behaviour is manifested, besides those political actions directly falling into the political arena, such as overtly addressing political matters and challenging the status quo, there are actions expressing political opinions in a more implicit or indirect way, such as the emergence of political consumerism (Zhang, 2015), environmentalism (Yang and Calhoun, 2007) and online political parody (Meng, 2011).

Based on these distinctions, four types of political engagement are identified, as indicated in Table 7.1, namely: (1) elite-challenging explicit participation; (2) elite-challenging implicit participation; (3) elite-directed explicit participation; and (4) elite-directed implicit participation.

Four types of political engagement

Elite-challenging manifested participation refers to all public and open resistance against the ruling authorities or policies, such as large-scale demonstrations, protests and collective actions (Cai, 2010). From 1993 to 2005, official reports of acts of contentious political actions rocketed from 8700 to 87 000, with many involving disruptive tactics or even violence (O'Brien and Stern, 2008). As reviewed by Lin and Zhang (2018), even though those elite-challenging collective actions in mainland China aim at advancing the public good, for fear of strong censorship and repression from the state, participants tend to avoid being formally organized, and online media become the major channel in the mobilization process. New media provide channels for Chinese netizens to participate, ranging from political discussion (Mou et al., 2011) to engaging in political activities, such as signing petitions and voting and polling online (Zhang and Lin, 2014).

The implicit elite-challenging participation – as an emerging way for citizens to engage in politics without directly challenging the status quo – jointly results from the state repression of elite-challenging behaviours and participants' active use of online media as a venue for public expression and resistance in an authoritarian context. This kind of political engagement takes various forms in both attitudinal and behavioural aspects. Attitudinally, several studies have documented how Chinese citizens cognitively engage in politics as a form of opposition against the dominating ruling principles, such as the formation of democratic knowledge and norms (Lei, 2011), and as a demand for more Internet freedom (Shen, 2017). In a 20-country comparative study, Shen (2017) found that, while mainland China is one of the most severely censored countries in the world, Chinese people reported a very high demand for Internet freedom, only second to Indonesia. Behaviourally, implicit elite-challenging participation appears in such forms as online parodies (Meng, 2011; Tang and Yang, 2011).

Online parodies, also called *e-gao*, or online spoofs, have become increasingly popular forms of political expression. They include various types of audio, visual or textual spoofs that are circulated on online platforms (Meng, 2011). Online parody is one example of how elements of entertainment, politics and popular culture can be combined, and it represents one of the many 'innovative strategies for articulating social critique and fostering societal dialogue in a heavily controlled speech environment' (Meng, 2011: 46). An example of online parody is one against the Green Dam Youth Escort (GDYE) software package (Lin et al., 2014). GDYE was required by the China Ministry of

Industry and Information Technology to be installed on all computers sold in China starting from 1 July 2009. The package would automatically censor websites deemed to be pornographic or subversive to the state. The policy triggered strong resistance from the public. Online protests, together with many other forms of parodies expressing discomfort with the GDYE and opposition to Internet censorship, resulted in a rare call-back of the policy.

Another type of political engagement in mainland China is environmentalism. Environmentalism is defined as the attitudes and behaviours that support the environmental movement and environmental protection (Stanley et al., 2017), such as the willingness to pay more money or support the government in increasing taxes to bolster pro-environmental activities. In mainland China, environment-related actions have strong political implications for two reasons. First, one of numerous types of civic actions, most environment-related activities share many of the same principles as movements concerned with human rights and peace (Schofer and Fourcade-Gourinchas, 2001). Most political activists conveniently used the environmental issue as an empirical case to express political concerns with governmental officials as the target of public grievances (Lin and Zhang, 2018). Environmental protests led to half of the 'mass incidents' that mobilized 10 000 or more participants between 2000 and 2013 (Li et al., 2014). Second, as important players in a civic society, environmental organizations and new media are often co-evolving (Yang, 2003a), such that the Internet facilitates civic activities by providing new possibilities for participation, and civic organizations promote the development of the Internet by providing necessary social bases for communication. Environmental organizations have been actively using the Internet to mobilize people to join in environmental protection behaviours (Hsu, 2010). Such online actions are forming a virtual 'green public sphere', which constitutes a foundation for further civic and political actions (Yang and Calhoun, 2007).

On the other hand, elite-directed political engagement in mainland China occurs when the ruling authorities intervene and regulate the procedures and scope of political participation, using it most often as a means of reinforcing the ruling party's legitimacy. Manifested political engagement directed by elites includes county-level elections – perhaps the only direct elections in which a Chinese citizen may participate (Zhong and Chen, 2002) – as well as various types of political campaigns initiated by the ruling party (Zhang and Lin, 2014). Different levels of government officials also use Internet-based platforms to facilitate public engagement and for citizens to voice their concerns, ranging from civic construction issues to social issues (Jiang et al., 2019).

Finally, the implicit elite-directed participation can be regarded as the routinized political campaigns initiated by the ruling authorities. For example, the government-organized non-governmental organizations (or GONGOs) have emerged to become visible players in Chinese environmental politics since the 1990s. They are set up in response to practical needs as well as the Chinese government's fear of bottom-up social mobilization on environmental issues (e.g., Dai et al., 2017; Wu, 2002). Another example in the online setting would be a government-launched application called 'Xuexi Qiangguo' (literally, 'a powerful learning country'). Users (essentially all party members) are required to sign up with their real names on a daily basis, and to earn points by reading articles, making comments and participating in multiple-choice tests about the party's policies (Huang, 2019). Although users performed interactions related to politics on the app, such as reading political news and expressing comments on the government and politics, those activities are designed to promote the ruling legitimacy, thus being far removed from the essence of political engagement designed to advance the public good.

Digital media uses and their political implications

In examining the media–politics linkage, previous studies conducted in mainland China have documented different usage dimensions of digital media, especially social media, and how they are related to different types of political engagement. For example, using social media for 'public affairs communication' – such as 'communication about issues and policies concerning the interests of the public' – is distinguished from 'communications about private lives, leisure activities, and/or entertainment' (Li et al., 2016: 5148). Echoing similar studies in the Western context (e.g., Moy et al., 2005), Zhang and Lin (2014) argued that social media usage is multidimensional. Specifically, four dimensions of social media use are: (1) information exchange and instrumental use; (2) relational and social networking use; (3) recreational or entertainment use; and (4) social media-based political activities.

In terms of the relationships between social media use and political engagement, previous work conducted in mainland China generally showed similar findings to what has been discovered in the West. For example, news and informational use of social media was found to positively relate to engagement in politics, whereas use of social media for entertainment was negatively related to political engagement (Zhang and Lin, 2018). Yang (2003b) found that social uses of the Internet would promote public debates and articulation of environmental problems, whereas the Internet helped traditional social organizations to build up virtual communities by increasing their publicity, disseminating information and networking with other international organizations.

With regard to the use of social media for entertainment, which used to be considered as irrelevant to politics, Yang found the opposite. It was found to have political implications, albeit contingent on the characteristics of individuals. For people with a high level of political trust and a high level of internal political efficacy, the recreational use of social media produced a positive political effect (Zhang and Lin, 2018). It is likely that entertainment media present foreign images and languages, as well as various cultural and political values, leading to politically meaningful messages. Specifically, online parody contains 'a system of coded language that defies immediate and complete control' when users communicate sensitive political issues via the Internet (Shen and Guo, 2013: 147). Those messages by themselves are a part of elite-challenging, implicit political engagement. One example was the use of the term 'river crab', a homonym for 'harmony' in Mandarin Chinese. Images of the river crab were used as a way of criticizing Internet censorship (Yang, 2016).

Moreover, unlike the West where netizens enjoy greater online freedom, the ostensive Internet censorship in China has become the backdrop of two lines of studies: first, research that analyses the use of online circumvention tools (*fan qiang*) to access the information that is filtered by the Great Firewall set up by the government; and second, research that analyses the responses towards the censorship itself. For instance, recent studies documented the extent to which Chinese netizens used circumvention tools to bypass the censorship (Mou et al., 2016; Shen and Zhang, 2018). From a nationally representative sample, Shen and Zhang (2018) found that more than 11 per cent of Chinese netizens had used circumvention tools at some time. Those users tended to be young with higher educational levels. In addition, those users were more likely to oppose censorship, have less trust in news media and participate more actively in civic activities, as compared to non-users. Zhong et al. (2017) reported people's responses to Internet censorship, such that perceived Internet censorship was negatively associated with the willingness to talk about sensitive issues and the likelihood of signing petitions with real names.

Finally, Skoric et al. (2016) found in their extensive meta-analysis that the expressive use of social media had a moderately strong relationship with political participation, but that the effects of positive relationships with informational and relational uses of social media on political participation were small. Previous studies also discovered indirect effects (via the mediation of interpersonal communication channels) and conditional (via the moderation of political trust and efficacy) media effects. For instance, Zhang and Lin (2018) found that social media use produced conditional effects on political participation. Informed by the communication mediation model, Zhong and Zhang (2017) revealed the pathways from digital media use to civic engagement, wherein civic discussion, political efficacy and emotions played a mediating role.

An agenda for future studies on digital media and politics in mainland China

Given the rapidly changing media and political landscape in mainland China, as well as several theoretical and methodological flaws in existing studies, this chapter raises three lines of inquiry for future studies. First, theme-wise, future studies may well consider focusing on how advanced information and communication technologies are involved in influencing people's political behaviour. Shorey and Howard (2016: 5042) proposed that China has become an important research area because its information infrastructure 'will shape the lives of a billion people . . . and the authorities are the source of algorithmic manipulations – such as social media bots that have an impact on public life in democracies'. The government has even been planning to implement the Social Credit System, which will rank the entire 1.4 billion Chinese citizens' social credit levels based on a comprehensive tracking and monitoring system (Liang et al., 2018). The system is 'a big data-enabled surveillance infrastructure to manage, monitor, and predict the trustworthiness of citizens, firms, organizations, and governments in China' (Liang et al., 2018: 415). Issues of press freedom, privacy, computational propaganda, mass surveillance and an informed citizenship are consequently coming to the fore. Further studies may examine the ways in which Chinese citizens interact with, as well as resist, the artificial intelligence (AI) technologies that are to be used to surveil and promote ruling authorities.

Second, when there is a strong presence of censorship and interference from the state, conventional political actions – especially those common in democratic societies, such as protests, petitions and other elite-challenging actions – are less likely to take place. Hence, future studies in the context of China might consider a focus on more nuanced and subtle social activities that nevertheless have political meaning and implications, such as in shaping the formation of alternative political beliefs.

Scholars have proposed the concept 'authoritarian deliberation' (He and Warren, 2011; Jiang, 2010) to explicate the fact that, even though the party-state limits and prescribes the boundaries of political discourses, the online deliberation of local collective problems – environmental issues, health issues, social services, among other topics – are somehow tolerated by the state in order to 'pacify the public and maintain government legitimacy . . . [and] to force the government, especially local governments, to be more efficient and accountable for their actions' (Jiang, 2010). Further studies are needed to examine the extent to which participatory behaviours within the party-state's boundary represent 'a worthwhile route to China's political reforms' (Jiang, 2010).

In fact, resistance from ordinary citizens may take place in non-political or even non-civic areas, such as resistance to the dominant lifestyle (Cui and Zhang, 2017), or the use of cultural and entertainment messages to voice their political concerns (Yang and Jiang, 2015). More studies are warranted. Furthermore, given the popularity of networked activities wherein users source and distribute their own information, future studies are called for to investigate the potential for a 'Fifth Estate' (Dutton, 2009; Huan et al., 2013) to emerge in this authoritarian context. In other words, the fact that new media users actively source their own information and create original content in various forms, such as microblogs, emails, tweets and comments, might as well provide a 'greater independence from other institutions and offer a mechanism whereby public opinion can be directly expressed' (Newman et al., 2012: 7).

Third, methodologically, this chapter calls for more robust research designs. Many existing studies have been heavily biased towards the younger generation, student samples, and the more developed large cities. In fact, the huge geographical span of China, with the cultural and language differences that exist within the mainland, requires more within-country studies from a comparative perspective. There is also an absence of media use and political participation studies targeting middle-aged or senior citizens, with some important exceptions (e.g., Xie, 2008). Studies based on nationally representative samples have often had crude measurements of political outcomes and media variables. Future studies would benefit from surveying wider populations, with samples that can support the study of different age groups, generational cohorts, ethnicities and language communities.

Furthermore, many studies rely on cross-sectional survey methods, which may suffer from self-reported biases, such as social desirability biases, particularly in such an authoritarian context. Hence, future research may consider collecting more behavioural data, such as trace data from social media, and implementing longitudinal research designs, such as panel survey studies. In 2010, a review of research argued that the 'lack of data is the biggest bottleneck for social scientists in China' (Zhu and Li, 2010). This bottleneck is opening up.

In addition, to establish causality, an experimental design or a mixed-methods approach might be adopted (Muise and Pan, 2019). Moreover, with the popularity of computational methods (that is, big data analytics, large-scale text processing, social network analysis, online field experiments, and agent-based modelling), researchers should consider the use of digital trace data or simulation to address focal research questions. With the continuous development of digital media in this regime, more theory-driven research with methodological innovations could shed new light on digital politics in mainland China.

References

Cai, Y. (2010) *Collective Resistance in China: Why Popular Protests Succeed or Fail.* Stanford, CA: Stanford University Press.

Cui, L. and Zhang, X. (2017) 'What Happened to Those Fans Several Years Later? Empowerment from Super Girls' Voice for Girls in China (2007–2015)', *Critical Studies in Media Communication,* 34(4), pp. 400–414.

Dai, J., Zeng, F. and Wang, Y. (2017) 'Publicity Strategies and Media Logic: Communication Campaigns of Environmental NGOs in China', *Chinese Journal of Communication,* 10(1), pp. 38–53.

Dutton, W.H. (2009) 'The Fifth Estate Emerging through the Network of Networks', *Prometheus,* 27(1), pp. 1–15.

Freedom House (2019) *Freedom of the World Report, 2018.* Available at https://freedomhouse.org/report/freedom-world/2019/china/ (accessed 1 September 2019).

He, B. and Warren, M.E. (2011) 'Authoritarian Deliberation: The Deliberative Turn in Chinese Political Development', *Perspectives on Politics,* 9(2), pp. 269–89.

Hsu, C. (2010) 'Beyond Civil Society: An Organizational Perspective on State–NGO Relations in the People's Republic of China', *Journal of Civil Society,* 6(3), pp. 259–77.

Huan, S., Dutton, W.H. and Shen, W. (2013) 'The Semi-Sovereign Netizen: The Fifth Estate in China'. In P.G. Nixon, R. Rawal and D. Mercea (eds), *Politics and the Internet in Comparative Context: Views From the Cloud.* London: Routledge, pp. 43–58.

Huang, Z. (2019) 'China's Most Popular App is a Propaganda Tool Teaching Xi Jinping Thought'. *South China Morning Post,* 14 February. Available at https://www.scmp.com/tech/apps-social/article/2186037/chinas-most-popular-app-propaganda-tool-teaching-xi-jinping-thought/ (accessed 1 September 2019).

Inglehart, R. (1977) *The Silent Revolution.* Princeton, NJ: Princeton University Press.

Jiang, J., Meng, T. and Zhang, Q. (2019) 'From Internet to Social Safety Net: The Policy Consequences of Online Participation in China', *Governance.* Available at https://onlinelibrary.wiley.com/doi/abs/10.1111/gove.12391/ (accessed 1 September 2019).

Jiang, M. (2010) 'Authoritarian Deliberation on Chinese Internet', *Electronic Journal of Communication,* 20(3-4). Available at https://ssrn.com/abstract=1439354/ (accessed 1 September 2019).

Kuan, H.C. and Lau, S.K. (2002) 'Traditional Orientations and Political Participation in Three Chinese Societies', *Journal of Contemporary China,* 11(31), pp. 297–318.

Lei, Y.W. (2011) 'The Political Consequences of the Rise of the Internet: Political Beliefs and Practices of Chinese Netizens', *Political Communication*, 28(3), pp. 291–322.

Li, L., Tian, H. and Lu, Y. (2014) *Zhongguo Fazhi Fazhan Baogao* (Annual Report on the Development of China's Rule of Law). Beijing: Chinese Academy of Social Science Press.

Li, X., Lee, F.L. and Li, Y. (2016) 'The Dual Impact of Social Media Under Networked Authoritarianism: Social Media Use, Civic Attitudes, and System Support in China', *International Journal of Communication*, 10, pp. 5143–63.

Liang, F., Das, V., Kostyuk, N. and Hussain, M.M. (2018) 'Constructing a Data-Driven Society: China's Social Credit System as a State Surveillance Infrastructure', *Policy and Internet*, 10(4), pp. 415–53.

Lin, F., Chang, T.-K. and Zhang, X. (2014) 'After the Spillover Effect: News Flows and Power Relations in Chinese Mainstream Media', *Asian Journal of Communication*, 25(3), pp. 235–54. doi: 10.1080/01292986.2014.955859.

Lin, F. and Zhang, X. (2018) 'Movement-Press Dynamics and News Diffusion: A Typology of Activism in Digital China', *China Review*, 18(2), pp. 33–63. Available at muse.jhu.edu/article/696528/.

Meng, B. (2011) 'From Steamed Bun to Grass Mud Horse: E Gao As Alternative Political Discourse on the Chinese Internet', *Global Media and Communication*, 7(1), pp. 33–51.

Mou, Y., Atkin, D. and Fu, H. (2011) 'Predicting Political Discussion in a Censored Online Environment', *Political Communication*, 28(3), pp. 341–56.

Mou, Y., Wu, K. and Atkin, D. (2016) 'Understanding the Use of Circumvention Tools to Bypass Online Censorship', *New Media and Society*, 8(5), pp. 837–56.

Moy, P., Manosevitch, E., Stamm, K. and Dunsmore, K. (2005) 'Linking Dimensions of Internet Use and Civic Engagement', *Journalism and Mass Communication Quarterly*, 82(3), pp. 571–86.

Muise, D. and Pan, J. (2019) 'Online Field Experiments', *Asian Journal of Communication*, 29(3), pp. 217–34.

Newman, N., Dutton, W.H. and Blank, G. (2012) 'Social Media in the Changing Ecology of News: The Fourth and Fifth Estates in Britain', *International Journal of Internet Science*, 7(1), pp. 6–22.

O'Brien, K.J. and Stern, R. (2008) 'Introduction: Studying Contention in Contemporary China'. In K.J. O'Brien (ed.), *Popular Protest in China*. Cambridge, MA: Harvard University Press, pp. 11–25.

Schofer, E. and Fourcade-Gourinchas, M. (2001) 'The Structural Contexts of Civic Engagement: Voluntary Association Membership in Comparative Perspective', *American Sociological Review*, 66(6), pp. 806–28. doi: 10.2307/3088874.

Shen, F. (2017) 'Internet Use, Freedom Supply, and Demand for Internet Freedom: A Cross-National Study of 20 Countries', *International Journal of Communication*, 11, pp. 2093–114.

Shen, F. and Guo, Z.S. (2013) 'The Last Refuge of Media Persuasion: News Use, National Pride and Political Trust in China', *Asian Journal of Communication*, 23(2), pp. 135–51.

Shen, F. and Zhang, Z. (2018) 'Do Circumvention Tools Promote Democratic Values? Exploring the Correlates of Anti-Censorship Technology Adoption in China', *Journal of Information Technology and Politics*, 15(2), pp. 1–16.

Shorey, S. and Howard, P. (2016) 'Automation, Big Data and Politics: A Research Review', *International Journal of Communication*, 10, pp. 5032–55.

Skoric, M.M., Zhu, Q. and Pang, N. (2016) 'Social Media, Political Expression, and Participation in Confucian Asia', *Chinese Journal of Communication*, 9(4), pp. 331–47.

Stanley, S.K., Wilson, M.S., Sibley, C.G. and Milfont, T.L. (2017) 'Dimensions of Social Dominance and Their Associations with Environmentalism', *Personality and Individual Differences*, 107, pp. 228–36. https://doi.org/10.1016/j.paid.2016.11.051/.

Tang, L. and Yang, P. (2011) 'Symbolic Power and the Internet: The Power of a "Horse"', *Media, Culture and Society*, 33(5), pp. 675–91.

Verba, S., Schlozman, K.L. and Brady, H.E. (1995) *Voice and Equality: Civic Voluntarism in American Politics*. Cambridge, MA: Harvard University Press.

Wu, F. (2002) 'New Partners or Old Brothers? GONGOs in Transnational Environmental Advocacy in China', *China Environment Series*, 5, pp. 45–58.

Xie, B. (2008) 'Civic Engagement among Older Chinese Internet Users', *Journal of Applied Gerontology*, 27(4), pp. 424–45.

Yang, F. (2016) 'Rethinking China's Internet Censorship: The Practice of Recoding and the Politics of Visibility', *New Media and Society*, 18(7), pp. 1364–81.

Yang, G. (2003a) 'The Co-Evolution of the Internet and Civil Society in China', *Asian Survey*, 43(3), pp. 405–22.

Yang, G. (2003b) 'The Internet and Civil Society in China: A Preliminary Assessment', *Journal of Contemporary China*, 12(36), pp. 453–75.

Yang, G. and Calhoun, C. (2007) 'Media, Civil Society, and the Rise of A Green Public Sphere in China', *China Information*, 21(2), pp. 211–36.

Yang, G. and Jiang, M. (2015) The Networked Practice of Online Political Satire in China: Between Ritual and Resistance', *International Communication Gazette*, 77(3), pp. 215–31.

Zhang, X. (2015) 'Voting with Dollars: A Cross-Polity and Multilevel Analysis of Political Consumerism', *International Journal of Consumer Studies*, 39(5), pp. 422–36. doi: 10.1111/ijcs.12181.

Zhang, X. and Lin, W.-Y. (2014) 'Political Participation in an Unlikely Place: How Individuals Engage in Politics through Social Networking Sites in China', *International Journal of Communication*, 8, pp. 21–42.

Zhang, X. and Lin, W.-Y. (2018) 'Stoking the Fires of Participation: Extending the Gamson Hypothesis on Social Media Use and Elite-Challenging Political Engagement', *Computers in Human Behavior*, 79, pp. 217–26. doi: 10.1016/j.chb.2017.10.036.

Zhong, Y. and Chen, J. (2002) 'To Vote or Not To Vote: An Analysis of Peasants' Participation in Chinese Village Elections', *Comparative Political Studies*, 35(6), pp. 686–712.

Zhong, Z.-J., Wang, T. and Huang, M. (2017) 'Does the Great Fire Wall Cause Self-Censorship? The Effects of Perceived Internet Regulation and the Justification of Regulation', *Internet Research*, 27(4), pp. 974–90.

Zhong, Z.-J. and Zhang, X. (2017) 'A Mediation Path to Chinese Netizens' Civic Engagement: The Effects of News Usage, Civic Motivations, Online Expression and Discussion', *China: An International Journal*, 15(2), pp. 22–43.

Zhu, J.H. and Li, X. (2010) 'Chinese e-Social Science: A Low-End Approach'. In W.H. Dutton and P.W. Jeffreys (eds), *World Wide Research*. Cambridge, MA: MIT Press, pp. 188–90.

PART III

Institutional Transformations

8 The politics of digital age governance

Volker Schneider

Introduction

The Digital Age, only half a century old, has shaped social, economic and political development in a profound way, and will transform it much more radically during the next few decades. It emerged in the 1960s with the use of computers in business and the armed forces, and spread to further sectors in the following decades. Only recently, it attained government and public administration. As in earlier ages, when new basic technologies produced new means of production and social organization, digital technology is also becoming the basis for new activities and organizations, which fundamentally change the political organization of society and its governance structures.

This chapter deals with these latter changes brought about by digital change. The central question is how both the inputs and outputs, and also the 'within-puts', of governing processes in modern societies have changed. To analyse these changes, an extensive concept of governance is used, which includes not only government and administration, but also civil society and the private sector, and multiple forms of socio-political participation and coordination. Before analysing and interpreting these changes at different levels and areas, the next section first presents the specific theoretical perspective used to analyse governance processes. In a concluding section, the main analytical results are briefly summarized.

Governance and politics in complex societies

Writing about politics and governance in the new technological context requires, first of all, a clear specification of both concepts. 'Governance' is one of the greatest conceptual innovations in the social sciences of recent decades (Weiss, 2000 gives a good conceptual overview). The term derives from 'govern' and 'government'. In the nominalized form of the verb 'to govern', this term extends its scope and makes it possible to decouple the function and the action set of governing from the institution of government, and to link it to a whole spectrum of governmental and non-governmental institutions, institutional arrangements and action sets which participate in the process of governing. Although the term is not new, it has only been popular since the 1990s, when political scientists tried to grasp coordination and cooperation arrangements at the global level (Rosenau and Czempiel, 1992), and institutional economists tried to understand the emergence of organizational hierarchies in contrast to market-based contractual systems (Williamson, 1991). The term has evolved to refer generally to the set-up and implementation of rule-based mechanisms to coordinate social and political interactions and to manage inherent socio-political conflicts.

However, limiting the term to rule-making and implementation would be too narrow. In principle, 'governance' includes all governmental and non-governmental activities and arrangements that are covered by the term 'policy-making'; that is, it also includes distributive, redistributive and persuasive activities in which public goods are produced and made available to society at large. In this sense it refers to an expanded perspective of policy-making in which a whole spectrum of actors and arrangements are included, from public to private, and from the local to the global level (Mayntz, 2003).

This perspective (see Schneider, 2012, 2020) makes it possible to distinguish at least four dimensions in societal differentiation that are relevant for governance arrangements:

- Institutional sectors partitioning government, civil society and the private market economy. Due to different institutional rules and resources, each of these sectors has different rights and capacities to act.
- Forms of relational coordination between hierarchy, networks and markets. Different configurations of exchange and control imply different mechanisms of coordination and cooperation.
- Functional differentiation structuring societies into multiple subsystems based on different communication rationalities and specialized expertise. Politics, economics, science, health, media, education, and so on imply not

only specific orientations and knowledge resources, but also specific organizational species and ecologies in which these orientations are routinized.
• Territorial levels of governance systems at the local, national, European or global level. These levels are often interdependent in configurations of multi-level governance. In climate policy, for example, urban governance interacts with areas of global governance.

In addition, a crucial facet of modern society is its interpenetration by large-scale socio-technical systems distributing energy, enabling transport and communication. Most of these systems imply specific governance arrangements (Schneider, 1991).

This multidimensional and multi-level perspective is particularly fruitful for the analysis of the complex societies of our time. It allows, for example, not only comparisons of how science, health care or energy systems are 'governed' at the national or global level, but also how specific governance structures differ in terms of impact and performance (Williamson, 1991). Comparative studies can identify advantageous and disadvantageous institutional arrangements. For instance, structures may be compared in which exclusively governmental organizations are involved, versus structures in which civil society and private actors are also contributing with their specific resources.

The governance perspective, however, has a crucial weakness with its focus on problem-solving. The issue is that in most cases, problem-solving is overdetermined by politics, that is, political conflicts over the acquisition and maintenance of power. Power in this perspective is not only limited to governments, but non-governmental actors such as companies, interest groups and social movements have control over power sources too, and are geared to defend them.

The unique position of governments – that is, the state – however, is that it is the only institution holding a monopoly of legitimate violence (Mann, 1984). Political theory explains the emergence of this centre of power either by outcomes of power struggles or by social contracts. However, concentration and centralization have always enabled the abuse of power, and societies have learned through painful experiences how to prevent these abuses by complex institutional designs. The polity of a country thus represents in some way the 'ossified outcomes' of historical power struggles and learning processes which often result in complex patterns of power-sharing (Schneider, 2015).

From this perspective, collaborative arrangements by policy networks should be regarded as modern forms of governance in which not only legislative, gov-

ernmental and administrative actors participate in the formulation and implementation of policies, but also private and civil society organizations from
the various societal subsystems contribute with their expertise and special
resources to the solving of societal problems (Kenis and Schneider, 2019).
Access to and participation in policy formulation and implementation ultimately implies that actors participate in the sharing of political power. In this
respect, it is important to pay attention to the specific institutional structures
that regulate access to these power positions and the many aspects of power
use. Since power positions imply privileges and comparative political advan-
tages in an 'ecology of policy games' (Dutton et al., 2012), the various players
are thus not just 'policy-seekers' as a pure governance perspective might
suggest, but also hunters for offices and political rents. Policies are sometimes
mere by-products of politics, when political actors are not genuinely interested
in the solving of a problem but just try, as policy entrepreneurs, to generate
political capital by the promotion of a popular policy topic. Governance is
thus always conditioned by the acquisition and retention of social and political
power.

Digitalization and the transformation of governance

The inclusive perspective on governance outlined above has important implications for the guiding question of this chapter. How digital technology
influences governance is thus not just an e-government issue, but a more
overarching problematique, as to how contemporary governance arrangements will reconfigure themselves in the course of digitization. The question
is not only how the inputs, within-puts and outputs of governmental action in
policy-making are changing in the course of digitization. It is also important
to know what changes are taking place in the non-state or private sectors, and
what shifts in position can be observed with respect to the relative prevalence
of coordination systems.

Digital technology and the growth of organizations

As a first step, it is necessary to describe the various shifts and changes in
governance that have taken place in the last decades. In this respect, the most
important question is how the digital age has affected the internal organization
of the state and its relative position with respect to other actors and sectors of
society. How did the bases of its 'autonomous power' – which typically include
central authority, administrative hierarchy, monopoly on the legitimate use of
force, and territorial sovereignty – change (Mann, 1984)?

The internal changes of the state in the digital age that affect the first three aspects have been studied thoroughly by Dunleavy et al. (2006). Particular emphasis here was placed on decentralization and disaggregation processes, as well as the introduction of market-like relations and accounting systems from the private sector. This initially took place under the flag of New Public Management, in which the economization of the public sector was apparently exaggerated. In some countries this was corrected by recentralizing processes, which Dunleavy labelled as a new phase of 'digital era governance'. However, despite partial corrections, these new adjustments in the organization of government and administration do not alter the megatrend of decentralization on the basis of increasing specialization. Unlike in nature, where species extinction is a major issue, the number of organizational species is not decreasing, but increasing (Gill, 2002).

The position of the state can also be evaluated in a relational perspective. How has the power of the state changed vis-à-vis other societal actors? What is the degree of autonomy of the state vis-à-vis civil society and the private sector, and to what degree is it dependent on resources controlled by other relevant societal actors? Many of these relations have changed significantly. A central thesis here is that digitization had a positive effect, not only on the growth of hierarchies in the form of private and public organizations, but also on the growth and expansion of markets. Both effects were important drivers of globalization, since digitization drastically reduced transaction costs, or better and more inclusively called governance costs.

The supportive effect for hierarchies is most evident in the private sector with the development of large multinational corporations (MNCs). In recent decades, companies have emerged with a size and reach that was unimaginable 50 years ago. We only have to imagine how such economic empires were integrated on the basis of analogue organizational tools to understand the benefits of digital technologies (Latham and Sassen, 2009). Email, video conferencing, management information systems, integrated accounting and warehousing systems are only the most important ingredients of an information and communication technology (ICT) revolution, which was undoubtedly one of the central drivers of globalization in recent decades (for an early case study, see Schneider, 1994). It had been supported simultaneously by liberalization and privatization policies, for which MNCs were the strongest advocates. That these processes coincided with the advent of digital technology is no accident. While digital technology was not the only factor that stimulated the concentration of economic power, it certainly was a critical element, without which the coordination and integration of these large economic units would have been unthinkable.

Liberalization and privatization produced market expansion, and declining communications and transportation costs removed traditional barriers to firm expansion. Intensified competition in the world market led to a large number of mergers during these decades, which transformed many large firms into political superpowers. The transnational growth of firms weakened the position of the state in several respects (Schneider, 2004):

- As the size of firms increased, so also the impact of single investment decisions on national economies grew, and governments became more vulnerable to economic pressure by giant corporations.
- The structural dependence of governments on private business increased, because transnational firms have more options for industrial location.
- Massive financial amounts can be transferred within corporate networks within seconds from one corner of the world to another, often without leaving the boundaries of a firm; governments thus lose control of financial flows.
- Corporate communication networks and accounting systems make it easy to use internal pricing to shift profits to countries or regions with lower tax schemes, which undermines the tax base of governments.
- Large firms often become increasingly powerful political actors which participate in many national and international policy fora. This also tends to weaken the position of governments, as they lose their formerly exclusive role in international relations.

In abstract terms, the growth of 'private hierarchies' has also changed the network position of 'public hierarchies'. In national and global policy networks, governments are often confronted with firms that command much greater expertise and influence resources than any single organization of the public sector. A striking example of the government's dependence on industry expertise is its cooperation in the fight against cyber-crime (Schneider and Hyner, 2006). These changes are weakening the position of governments in general.

At the same time, a pluralization of organizational forms took place at the national and international level, and a proliferation of interorganizational networks in which different modes of 'network governance' (Provan and Kenis, 2008) emerged, that would be inconceivable without modern ICTs. Governance-cost-reducing effects of digital technologies can be observed in all forms of formal organizations that Coleman (1974) once theorized as 'corporate actors'. These range from public administrations and large companies to business associations, public interest groups, and civil society organizations representing social movements. For all these formal organizations, ICTs help to make both internal (integration and control) and external relationships (cooperation and resource mobilization) more effective and efficient.

Organizational growth can be observed particularly in international politics, in which policy processes are increasingly also shaped by non-governmental organizations. These include business associations, federations of parties, trade unions and global public interest groups. Social movements have also shifted organizational capacities to the global level, fighting for global environmental protection or the universal protection of human rights, to give some examples. Digital technologies are important drivers in this proliferation process. Just as digital technology has already promoted the spread of the anti-globalization movement ATTAC, so also the current Fridays for Future movement for climate protection would never have been able to gain a global reach in such a short time without email, websites, videoconferences and social media.

The growth and spread of national and international think tanks (Weaver, 2000), another rather new organizational species, is also strongly linked to digital technologies. The Internet not only opened access to scientific knowledge for research outside universities and established research institutes, but also facilitated the dissemination of research output to societal and political stakeholders. In recent decades, a huge spectrum of organizations and institutes emerged that try to influence policy formulation by generating and distributing scientific knowledge, ranging from large corporate think tanks to small groups of 'citizen scientists', thus supporting the trend to the scientification of policy-making.

Digital technology and the expansion of markets and hybrid forms of coordination and cooperation

Digitization influenced the growth of organizations and lowered the costs of hierarchical integration. At the same time, however, it also enabled the spread and expansion of market coordination into areas where competition was heretofore unsustainable. At the same time, digital technologies enabled new complex forms of economic organization, such as the various arrangements within the sharing economy which would not have been feasible with analogue technologies. Finally, digital technologies also created many new products that populate the information sector of today.

An important stimulus came by the retreat of the state from the provision of infrastructures since the 1980s, which opened new areas for private firms (Schneider et al., 2005). This was accompanied by 'de-politicization' of traditional regulatory structures and the global diffusion of independent regulatory agencies (Jordana et al., 2011).

Prior to this development, many government-provided infrastructural services were considered to be natural monopolies, that is, types of economic transactions where market coordination tends to fail. In such sectors, the problem was network externalities based on positive feedbacks of network size on returns and consumer utility, a mechanism that led everywhere to monopolization. Since private monopolies were considered to be socially unacceptable because of monopoly pricing and a disregard of infrastructural needs, the transformation of these systems into state-owned or publicly regulated monopolies was seen as a solution in most developed countries. Infrastructural systems such as telecommunications or energy thus were provided as nationalized or governmentally regulated systems. Other sectors such as broadcasting or road traffic were often considered to be public goods in the sense that it was too costly to exclude consumers from the use of these services. Because private systems could not be financed on this logic, the demand for such services had to be met through public provision.

Most of these effects have been mitigated or abolished by the advent of digital technology. Digitization has enabled not only the introduction of network competition and network interconnection, but also the introduction of cost-saving access systems. Services formerly displaying features of public goods could thus be transformed into private markets. New control technologies based on encoding and decoding devices facilitated the specification, assignment and enforcement of private property rights, to create new commercialized areas. An example is the extension of intellectual property by recent property protection and copyright laws and digital rights management systems. Formerly non-commercial domains, often controlled by governments or 'communities' (for example, science) where commercial transactions did not work because of technical or normative constraints, have thus been transformed into emerging markets.

New ICTs economized governance costs, so that market coordination even became possible in complex and interdependent networks. An example of the integration of a whole spectrum of coordination systems between market and hierarchy is provided by decentralized energy systems based on renewable energy. Another example is emissions trading systems for the purpose of climate protection. Both socio-technical systems would be unthinkable based on analogue technologies.

These changes, in which digital technologies became indispensable for the coordination and control of modern societies, were not purely techno-determined processes. Many of these changes were contested, and therefore subject to politics. Particular groups interested in market potentials were opposed by social and political groups defending traditional areas of public control. The fact that competition and market enlargements became technically feasible is certainly an effect of technological change. But how this potential is ultimately exploited depends a lot on political processes and power struggles. The different stages and outcomes of such battles varied between countries according to institutional conditions and situational power relations. These cross-national differences are particularly strong, for instance, in the digitization of health systems, or in the digitization of public administration. Federalist political structures or varying organizational capacities of interest groups seem to play an important role in the differential digitization of these sectors.

Digital technology and the repositioning of the state

The unprecedented growth of organizations and markets has undoubtedly also changed the position of the state in society and its dependence on the economy, although we can observe some interesting ambiguities. An important feature of societal development during the last century was the rise of formal organizations. In the second half of the twentieth century they reached sizes and scopes that exceeded that of many nation-states. The dominance of this social configuration has led Herbert Simon to call contemporary societies 'organizational economies' instead of 'market economies', despite the fact that the linkages between organizational hierarchies are usually networks of market relationships (Simon, 1991). Digital technologies have played an immense role in expansion of these market relations in recent decades. At least for all advanced industrial societies, the position of government and public administration in the social fabric has clearly lost its dominance and centrality vis-à-vis the economy and civil society.

However, an interesting development is that the predominance of large organizations over individual citizens has changed too, and digital technologies also played a role in this shift. Both via social media (in terms of agenda-setting) and via word of mouth (consumer-to-consumer communication about product and service quality), individuals have a much stronger position today, now that their voice is heard. Nowadays, a single tweet or an individual video posted by a blogger can trigger a government crisis or a corporate crash (Dutton, 2012). This means that digital technology can lead in some way to the restitution of individual power, reversing the process of power shifts from the individual to the corporate level that Coleman (1974) observed half a century ago.

How powerfully this spontaneous and unorganized area of the Internet will develop in future also depends on how governments or corporations – for instance, within arrangements of 'regulated self-regulation' as was designed by the German Hate Speech regulation – manage to get these dynamics under control (Kenis and Schneider, 2019).

Other digital technologies that strengthen both individual and private corporate actors are encryption techniques and closed sub-networks on the Internet, in which communications and transactions can flow uncontrolled. We can assume that within the coming information economy an increasing share of transactions may be shifted to closed, firewalled and even 'darknets' of the Internet. It also seems plausible that the potential of cryptology will increasingly be used in the form of digital money, and government will have difficulties in defending some of their central prerogatives: the monetary and fiscal sovereignty of the state.

It has been mentioned that some developments are ambivalent and contradictory. Some technologies strengthen, some weaken government and administration. There is a growing array of technologies that strengthen the power of the state to Orwellian dimensions. In recent decades, a variety of technologies have been developed – intelligent cameras, biometric identification, radio-frequency identification (RFID) chips, digital rights management systems, data mining and matching technologies, and so on – that provide law enforcement, security and military agencies with a whole new range of surveillance and control technologies. Without limits set by data and privacy protection regulations, the power of the state may be pushed to unprecedented heights. Never before in the history of the state have its agencies had such powerful tools to monitor, identify, categorize, track and record the communications, transactions and movements of its citizens. An increasing number of critical scholars fear the rise of the 'surveillance state'. A most extreme form is the current use of digital technologies for a social surveillance and credit system in China which is also combined with Internet censorship (Qiang, 2019).

From a purely technical perspective, the Digital Age therefore does not necessarily signify the erosion of statehood, as Cassandra-like prophecies in the 1990s have dramatized. In contrast, more plausible seems to be a kind of co-evolution in a form of an 'arms race' between state and society, where the state will certainly not lose out in the long run. At this level, digital technology as such is not the single determinant, but an important enabling and facilitating factor. If we recall Gordon Moore's Law, according to which computing power exponentially increases and computer prices decrease, tools of surveillance will become cheaper and more powerful, enabling governmental agencies to expand the breadth and depth of their surveillance activities. At the same time, digital technologies will also enable defensive technologies, and support collective action to politically contain some of these opportunities.

Conclusion

Observations and reflections on the effects of digital technologies on governance structures in modern societies have shown that while some basic trends can be observed and explained, the basic question of the possible repositioning of the state, as the core of the governance complex, can only be answered to a limited extent. One important reason is the contradictory developments outlined above, which in part weaken but also can strengthen the state. This ambiguous pattern of development could refer to a number of conflicting forces that Michael Mann (1984) had identified as dialectical patterns of state and social development: governmental organizations develop infrastructure power and extract resources for this function, and are also supported by civil society. Infrastructural investments promote digital technologies as communication and information infrastructures, which in turn promote organizations and markets, and also the growth and consolidation of civil society. As civil society gains in importance, the state is becoming increasingly dependent on these social sectors. At the same time, however, developments may occur in which civil society loses control over the state and its accumulating infrastructure power, and the state can develop despotic forms of power, as we are currently seeing in some totalitarian and neo-authoritarian regimes. But as digital infrastructures ultimately permeate society as a whole, the growth of markets and organizations may also strengthen civil society again, which can regain and defend its power in a further step. Whether the pluralistic, differentiated and networked governance structures of modern societies, which would not be functional without their digital infrastructures, are preserved as such and are not absorbed by totalitarian and neo-authoritarian developments, is ultimately not a question of technology but of socio-political power struggles.

References

Coleman, J.S. (1974). *Power and the Structure of Society*. New York: Norton.
Dunleavy, P., Margetts, H., Tinkler, J. and Bastow, S. (2006) *Digital Era Governance: IT Corporations, the State, and e-Government*. Oxford: Oxford University Press.

Dutton, W.H. (2012) 'The Fifth Estate: A New Governance Challenge'. In D. Levi-Faur (ed.), *The Oxford Handbook of Governance*. Oxford: Oxford University Press, pp. 584–98.

Dutton, W.H., Schneider, V. and Vedel, T. (2012) 'Large Technical Systems as Ecologies of Games: Cases from Telecommunications to the Internet'. In J. Bauer, A. Lang and V. Schneider (eds), *Innovation Policies and Governance in High-Technology Industries: The Complexity of Coordination*. Berlin: Springer, pp. 49–75.

Gill, D. (2002) 'Signposting the Zoo', *OECD Journal on Budgeting*, 2(1), pp. 27–79.

Jordana, J., Levi-Faur, D. and i Marín, X.F. (2011) 'The Global Diffusion of Regulatory Agencies: Channels of Transfer and Stages of Diffusion', *Comparative Political Studies*, 44(10), pp. 1343–69. doi:10.1177/0010414011407466.

Kenis, P. and Schneider, V. (2019) 'Analyzing Policy-Making II: Policy Network Analysis'. In H. van den Bulck, M. Puppis, K. Donders and L.v. Audenhove (eds), *Handbook of Media Policy Research Methods*. Basingstoke: Palgrave Macmillan, pp. 471–91.

Latham, R. and Sassen, S. (2009) *Digital Formations: IT and New Architectures in the Global Realm*. Princeton, NJ: Princeton University Press.

Mann, M. (1984) 'The Autonomous Power of the State: Its Origins, Mechanisms and Results', *European Journal of Sociology*, 25(2), pp. 185–213. doi:10.1017/S0003975600004239.

Mayntz, R. (2003) 'New Challenges to Governance Theory'. In H.P. Bang (ed.), *Governance as Social and Political Communication*. Manchester: Manchester University Press, pp. 27–40.

Provan, K.G. and Kenis, P. (2008) 'Modes of Network Governance: Structure, Management, and Effectiveness', *Journal of Public Administration Research and Theory*, 18(2), pp. 229–52.

Qiang, X. (2019) 'The Road to Digital Unfreedom: President Xi's Surveillance State', *Journal of Democracy*, 30(1), pp. 53–67.

Rosenau, J.N. and Czempiel, E.-O. (eds) (1992) *Governance without Government: Order and Change in World Politics*. Cambridge: Cambridge University Press.

Schneider, V. (1991) 'The Governance of Large Technical Systems: The Case of Telecommunications'. In T.R. La Porte (ed.), *Responding to Large Technical Systems: Control or Anticipation*. Dordrecht: Kluwer, pp. 18–40.

Schneider, V. (1994) 'Multinationals in Transition: Global Technical Integration and the Role of Corporate Telecommunication Networks'. In J. Summerton (ed.), *Large Technical Systems in Change*. Boulder, CO: Westview, pp. 71–92.

Schneider, V. (2004) 'The Transformation of the State in the Digital Age'. In S. Puntscher Riekmann, M. Mokre and M. Latzer (eds), *The State of Europe: Transformations of Statehood from a European Perspective*. Frankfurt/Main: Campus, pp. 51–72.

Schneider, V. (2012) 'Governance and Complexity'. In D. Levi-Faur (ed.), *Oxford Handbook on Governance*. Oxford: Oxford University Press, pp. 129–42.

Schneider, V. (2015) 'Towards Post-Democracy or Complex Power Sharing? Environmental Policy Networks in Germany'. In V. Schneider and B. Eberlein (eds), *Complex Democracy Varieties, Crises, and Transformations*. Berlin: Springer, pp. 263–79.

Schneider, V. (2020) 'Bringing Society Back in: Actors, Networks, and Systems in Public Policy'. In H. Lehtimäki, P. Uusikylä and A. Smedlund (eds), *Society as an Interaction Space: A Systemic Approach*. Berlin: Springer.

Schneider, V., Fink, S. and Tenbücken, M. (2005) 'Buying out the State: A Comparative Perspective on the Privatization of Infrastructures', *Comparative Political Studies*, 38, pp. 704–27.

Schneider, V. and Hyner, D. (2006) 'Security in Cyberspace: Governance by Transnational Policy Networks'. In M. Koenig-Archibugi and M. Zürn (eds), *New Modes of Governance in the Global System: Exploring Publicness, Delegation and Inclusiveness*. New York: Palgrave, pp. 154–76.

Simon, H.A. (1991) 'Organizations and Markets', *Journal of Economic Perspectives*, 5(2), 25–44. doi:10.1257/jep.5.2.25.

Weaver, R.K. ed. (2000). *Think Tanks and Civil Societies: Catalysts for Ideas and Action.* New York: Routledge.

Weiss, T.G. (2000) 'Governance, 'Good Governance and Global Governance: Conceptual and Actual Challenges', *Third World Quarterly*, 21(5), 795–814. doi:10 .1080/713701075.

Williamson, O.E. (1991) 'Comparative Economic Organization: The Analysis of Discrete Structural Alternatives', *Administrative Science Quarterly*, 36(2), 269–96. doi:10.2307/2393356.

9   How accountable are digital platforms?

Giles Moss and Heather Ford

Introduction

Recent debates about 'the digital' have been preoccupied with the power of a small number of platforms and their corporate owners: Amazon, Apple, Facebook, Google and Microsoft. Concerns about the power of media organizations are not new. There are long-standing debates about concentrated ownership in print media and broadcasting. Similar concerns did not emerge in the case of digital platforms because of an assumption that they were neutral and mere facilitators of different uses (Gillespie, 2010). As their neutrality has been challenged, as the way platforms shape our actions and relationships has become clear, questions about their power and responsibilities have followed.

A key question for researchers of digital politics, we argue here, is how platform power is held accountable. We set out a research agenda for answering this question that is both empirical and normative. On the one hand, we need to trace how accountability operates. What accountability mechanisms exist, how are they used by publics, how do platforms respond, and with what effects? At the same time, we need to reflect on what platform accountability means normatively. What does genuine accountability require, and how do existing accountability practices compare with this standard? Informed by deliberative approaches to democracy, and drawing in particular on Rainer Forst's (2014a, 2014b, 2017) work on justification, we argue that the accountability of platforms is a question of their power being justified adequately to affected publics, and that this depends on the quality of the discursive processes through which decisions about platforms are justified. Importantly, focusing on the quality of discursive processes allows us to distinguish critically between cases where publics merely 'accept' platform power unreflectively and in contexts of limited information and choice, and cases where power is justified through 'good' reasons tested through inclusive public discourse (Forst, 2017: 37–54).

Platform power, accountability and justification

There has been a shift towards the 'platformization of the Internet' in recent years, as a small number of platforms have come to dominate Internet use (Flew et al., 2019: 43). The term 'platform' is fuzzy, covering meanings from both computing and economics. As Flew et al. (2019: 36) explain, 'the contemporary concept of a platform brings together definitions from information engineering, where platforms are modular software architecture and economics, where they function as the meeting place for double-sided markets'. Van Dijck et al. (2018: 4) provide a useful definition: 'an online "platform" is a programmable digital architecture designed to organize interactions between users – not just end users, but also corporate entities and public bodies'. They explain how platforms are 'fuelled by *data*, automated and organized through *algorithms* and *interfaces*, formalized through *ownership* relations driven by *business models*, and governed through *user agreements*' (van Dijck et al., 2018: 9). Various online sites and services fall under this definition, but particular concerns have been expressed about the prominent platforms owned by companies such as Amazon, Apple, Facebook, Google and Microsoft.

Economists explain why a small number of platforms tend to dominate Internet use, despite the open and decentralized nature of the Internet. Barwise and Watkins (2018) outline four main reasons why digital markets are 'winner-takes-all': (1) economies of scale (resulting from high production costs and low distribution costs in digital content); (2) direct and indirect network effects (the value of a platform increases with additional users); (3) switching costs (users face significant costs when changing platforms); and (4) big data (platforms collect user data that can be analysed for competitive advantage). These factors combine to benefit established platforms, enabling them to maintain a dominant position over would-be competitors. Companies may also bolster their position by acquiring emerging platforms (for example, Facebook's acquisition of Instagram in 2012 and WhatsApp in 2014).

The implications of economically dominant platforms go beyond the economic. While platforms allow us to do a range of things conveniently, from staying in touch with others to purchasing new products, they are not just facilitators. Platforms shape our actions and relationships in particular ways (van Dijck et al., 2018: 8). As such, platform owners may exercise 'power':

'the capacity of A to motivate B to think or do something that B would not otherwise have thought or done' (Forst, 2017: 40). Whether through human or automated means, platforms decide what content we receive, determining what is acceptable and unacceptable and what is promoted or not (Gillespie, 2018). The fine-grained data platforms have about their users may be exploited to 'nudge' user decisions in particular directions, for commercial reasons or even to try and influence election outcomes (Tambini, 2018; Yeung, 2017). Since platforms penetrate our cultural, economic and political lives so deeply, the consequences of their power are far-reaching.

More research is required to understand platform power in all its complexity; we emphasize the need for transparency to facilitate such research below. Our focus here is on how this power is held accountable. Insofar as platforms can exercise power, there is the possibility that it may be misused. As Baker (2007: 16) explains, 'Concentrated media ownership creates the possibility of an individual decision maker exercising enormous, unequal and hence undemocratic, largely unchecked, potentially irresponsible power'. He argues: 'Even if this power were seldom if ever exercised, the democratic safeguard value amounts to an assertion that no democracy should risk the danger' (ibid.). Understood this way, the issue is not so much how platform power is currently used by platform owners, but what mechanisms, if any, affected publics can use to hold platforms to account if required.

Evaluating accountability is not straightforward. On the one hand, the fact that platform owners may resist particular public demands does not mean that accountability is absent. Being accountable does not mean that platforms always change decisions when faced with public pressure, especially because publics often disagree about what actions are appropriate for platforms to take. On the other hand, the fact that the public may not resist particular decisions does not mean that accountability is present. Publics might lack adequate information about platform governance, or meaningful opportunities to reflect on its consequences. Even if informed, they may feel 'powerless' (Andrejevic, 2014) or 'resigned' to the current situation (Draper and Turow, 2019), believing there is 'no alternative' (Dencik, 2018).

Accountability is closely tied to justification. As Giddens (1984: 30) explains, 'to be "accountable" for one's activities is both to explicate the reasons for them and supply normative grounds whereby they can be "justified"'. The key question, then, is how we can determine whether justifications are adequate. Informed by deliberative conceptions of democracy (Habermas, 1997; Mansbridge et al., 2012), and drawing in particular on the theory of justification developed by Forst (2014a, 2014b, 2017; see Moss, 2018), we argue that adequate justification depends on the quality of the processes of public discourse through which power is justified to the publics it may affect. In this view, a platform may be accountable even if it resists specific public demands, if platform owners provide reasons for their decisions in line with shared normative values and those affected have meaningful opportunities to reflect on, revise and reject these reasons (Thompson, 2016). At the same time, a platform may not be accountable even when its power is seemingly accepted by large numbers of users, if processes of justification are inadequate and those affected have limited information, choice and opportunity to reflect on the implications of platform governance. After all, as stated in the introduction, there is an important difference between power being merely 'accepted' by publics, unreflectively and in contexts of limited information or choice, and justifications that are based on 'good' reasons tested through inclusive public discourse (Forst, 2017: 37–54).

We outline an agenda for researching platform accountability below that is both empirical and normative. One task is to trace how accountability operates. What opportunities for accountability exist, how are they used by publics, how do platforms and other actors respond, and with what effects? A second task is to consider what genuine accountability requires, and to evaluate accountability practices against these standards. What is the quality of the discursive processes of justification between platforms and publics, and what criteria can we use to assess these processes?

Analysing accountability practices

There is an established literature on the accountability of print media and broadcasters, which provides a useful starting point for thinking about platform accountability (Bardoel and d'Haenens, 2004; McQuail, 2005; Eberwein et al., 2011). Drawing on this literature, and in particular McQuail (2005), we can identify four broad accountability mechanisms relevant to platforms: market, self-regulation, the public sphere and political representatives.

Market

As the idea of platform neutrality has been challenged, the claim that platform power is held accountable through the market has increasingly taken its place. Market accountability is based on the idea that competition between companies will ensure that they respond to consumer demand. Since dissatisfied consumers will find alternative providers, companies only succeed if they 'give consumers what they want', and so they must work hard to satisfy consumer preferences. From this perspective, the fact that people decide to use platforms indicates that they are getting what they want and that they support the way platforms are governed.

Market accountability requires competition. Without it, companies need not be responsive to consumer demand. As already noted, however, the 'winner-takes-all' nature of digital markets tends to weaken competition. Indeed, much of the current concern about platform power stems from a belief that particular platforms have become too large and dominant (Moore and Tambini, 2018). But the market is not the only accountability mechanism available to publics. In a well-known analysis, Hirschman (1970) argues that we can try to influence organizations through 'voice' as well as 'exit'. Voice involves potentially thicker forms of communication where publics raise concerns, either directly to platforms through the self-regulatory mechanisms platforms provide or indirectly through the public sphere and political representatives.

Self-regulation

The self-regulatory practices that platforms voluntarily adopt offer another accountability mechanism. As Crawford and Gillespie (2014: 3) note, 'social media platforms generally go well beyond what is legally required, forging their own space of responsibility'. Platforms communicate information to users about their governance and offer mechanisms for users to give feedback and raise concerns. In this way, the architectures of platforms generate their own system of regulation (Lessig, 2006), situating owners and users in ways that enable different levels of control.

Self-regulation mechanisms range from simple reporting tools to participatory discussion spaces. Crawford and Gillespie (2014: 413) discuss the limitations of the commonly used method of flagging: 'flags are a thin form of communication, remarkable more for what they cannot express than what they can'. They contrast flagging with the thicker communication facilitated by Wikipedia discussion pages, where 'the quality of content is openly debated and the decisions to keep or remove content on that basis are visible and preserved over time' (Crawford and Gillespie, 2014: 421). In principle, self-regulation has the advantage of being a flexible, extensive and participatory form of accountability (McQuail, 2005: 99). However, the commitment of platforms to effective self-regulation may be questioned. Platforms owned by private companies, answerable to advertisers and shareholders as well as users, may use self-regulation to forestall government regulation, but not in a way that achieves genuine accountability (Freedman, 2016).

Public sphere

The public sphere is another accountability mechanism, insofar as platforms may be influenced by and need to respond to broader public discourses about platform governance (Habermas, 1997: 329–88). Journalists, experts and various digital-rights groups scrutinize platforms and, at times, concerns about how platforms are governed have escalated into wider public controversies. Such controversies are significant since they may indicate the failure of other accountability mechanisms and may raise public awareness of particular aspects of platform governance.

Ananny and Gillespie (2017) consider the role that such public controversies play. As well as requiring platforms to justify decisions, controversies may bring changes in platform governance. However, they question whether public controversies, which tend to be episodic and focused on a specific problem with a particular platform, enable publics to respond meaningfully to longer-term, more systemic issues posed by platform power. There is also the critical question of whether concerns expressed in the public sphere are picked up by political representatives and addressed through effective policy.

Political representatives

The final accountability mechanism is provided by political representatives and policy-makers. Publics may hold platforms to account through political representatives in national governments or supranational organizations such as the European Union. The global nature of platforms poses governance problems. While platforms may be subject to the legal authority of governments in the territories in which they operate, and governments may intervene to shape platform governance, the capacity of different governments varies significantly. Furthermore, policy interventions might have impacts beyond specific territories, so international dialogues in spaces such as the United Nations' Internet Governance Forum, which brings together governments, the private sector and civil society groups, are especially important.

There is a range of potential policy responses to platform power. Policy can focus on platform content, seeking, for example, to restrict 'bad' content or promote 'good' content. Policy might focus instead on platform practices by, for example, ensuring that certain standards of privacy and data protection are met. Policy-makers may achieve accountability through law, rules overseen by regulatory authorities or through what Bunting (2018) calls 'procedural accountability', where regulators assess the self-regulatory processes that platforms adopt. Finally, policy might focus on platform structures. For example, large companies could be divided, and mergers and acquisitions prevented, to promote market competition. Most ambitiously, Andrejevic (2013) suggests that governments could support alternative non-commercial platforms, such as public service social networks or search engines.

Of course, to provide effective accountability, political representatives must be responsive to public concerns themselves. Within representative democracies, voting provides a fundamental but limited link between publics and elected representatives. But questions still emerge in formally democratic contexts about the extent to which all perspectives are represented in media policy-making (Freedman, 2008: 80–104). Beyond infrequent elections, we need to ask about the extent to which publics have meaningful opportunities to engage with, reflect on and influence policy outcomes.

While the four mechanisms discussed above provide a useful starting point, in-depth empirical research is required to understand how they operate. Firstly, research needs to understand how these mechanisms interact, to provide affected publics with accountability opportunities. Such research would be comparative. The opportunities for accountability that exist will vary between platforms, not only between private and public or non-commercial and largely volunteer-driven platforms, such as Wikipedia, but also among private platforms. We should also expect differences across countries, and that the opportunities available to publics will not be distributed evenly in different political contexts. As well as mapping accountability opportunities, research then needs to examine how these opportunities are taken up by publics in specific accountability practices in relation to particular issues. Through case study research, focused on tracing concrete practices, we can examine how different groups use the opportunities that are available, how platforms and other actors respond, and with what effects.

Evaluating relations of justification

Analysing how accountability practices operate is one thing; determining whether genuine accountability results from these practices is another. Evaluating accountability is not easy. As noted above, the fact that a platform may resist particular public demands does not mean that it is unaccountable. Furthermore, a lack of public resistance to their decisions does not mean that they are accountable. From a market-based perspective, it might be argued that the large numbers of people who choose to use platforms proves that platforms give users what they want and that platform power is therefore 'justified'. However, we may question whether platform power is justified in these circumstances if users lack adequate information, meaningful choice and opportunity to reflect on the broader implications of this power.

As we noted above, the accountability of platforms is a question of their power being justified adequately to affected publics. Adequate justification depends, in turn, on the quality of the process of public discourse by which reasons are given and those affected may reflect on, revise and reject these reasons; put differently, it depends on the quality of 'relations of justification' (Forst, 2014b: 1–14). Drawing on deliberative and participatory theories of democracy (Habermas, 1997; Mansbridge et al., 2012), and in particular the 'democratic goods' identified by Smith (2009),[1] we can identify four key criteria that can be used to evaluate the quality of relations of justification between platforms and affected publics: inclusion, transparency, deliberation and empowerment.

Inclusion

Accountability practices must be inclusive so that the perspectives of all affected are represented. Even if platform decisions are accepted by the majority, minority groups may be affected disproportionately and should have an opportunity to raise concerns. For example, DeNardis and Hackl (2016: 763) discuss the way platform governance may affect lesbian, gay, bisexual and transgender (LGBT) rights, and describe the impact of Facebook's real-name policy on drag communities. Platform power may also have consequences for non-users, as is the case, for example, with concerns about the potential impact of some platforms on electoral outcomes. Affected publics may therefore be wider than platform users. Where this is the case, non-users should be able to raise concerns; something which, of course, is more possible through some accountability mechanisms than others.

Transparency

Affected publics need clear and understandable information about how platforms are governed, including technical processes. Third parties (civil society organizations, governments, journalists and researchers) can help to promote public understanding, but their efforts may be hindered by limited access to relevant platform data. Where users lack relevant information, their seeming acceptance of platform power may be questioned. Yet the details of how platforms are governed can be opaque to interested observers, let alone other affected groups.

Deliberation

Publics require meaningful opportunities to reflect on the consequences of platform power. Platforms should justify their decisions in line with important normative values, such as fundamental rights. As Thompson (2016: 847) has argued, platforms should display 'normative integrity – a commitment to trying assiduously enough to succeed in understanding and evaluating the facts brought before them, in coherence with the central normative commitments of the communities they inhabit'. Normative values are not always reflected by the market, which may encourage the overproduction of 'negative external-ities' (for example, hateful discourses) and the underproduction of 'positive externalities' (for example, well-informed publics) (Baker, 2002: 41–62). If provided with opportunities to reflect deliberatively on normative values and consequences, publics may question the dominant justifications of platforms.

Accountability through 'voice' is important here, but as noted above, discourse in the public sphere can be episodic rather than sustained and systematic, while political processes may fail to engage and respond to publics adequately. Gillespie (2018: 210) urges platforms themselves to do more to facilitate the voice of users as citizens. He writes that, 'platforms should be developing structures for soliciting the opinions and judgement of users in the governance of groups, in the design of new features, in the moderation of content'. He suggests that platforms should be able to glean 'from users their civic com-mitments – not what they like as consumers, but what they value as citizens' (ibid.).

Empowerment

Publics need to be empowered with effective choice, whether through the market, platform self-regulation, or government policy. Andrejevic (2014: 1685) argues that users feel 'powerless' in relation to digital surveillance: 'People are palpably aware that powerful commercial interests shape the terms of access that extract information from them: they must choose either to accept the terms on offer or to go without resources that in many ways are treated as utilities of increasing importance in their personal and professional lives.' Likewise, Draper and Turow (2019) refer to people's 'digital resignation'. In circumstances where users have limited choice, their seeming acceptance of platform power may be questioned.

Where competition exists, users may exit and find alternative platforms to use. But, even in a competitive market, companies will only respond to what users want as consumers rather than citizens. Publics may be empowered in other ways. As we have stressed, platforms need not always shift position in the face of public pressure, but changes in platform governance as a result of public discourse should be possible, whether this is reflected in the policies of particular platforms, regulations governing platforms or through alternative public platforms. The range of policy interventions available to governments can offer meaningful choices that dispel the disempowering sense that 'there is no alternative' (Dencik, 2018).

The responsiveness of political representatives in picking up public concerns and resolving them effectively is crucial. As noted earlier, questions remain about the extent to which all perspectives are represented adequately in policy-making. Some of the gravest concerns about platform power surface here. Tambini (2018: 282) invites us to imagine the following:

[I]f a party or campaign emerged that was standing on a platform of breaking up social media companies, there would be a strong incentive for social media companies to undermine the visibility of that party. This example may, or may not be far-fetched, but parties already exist that propose radical, sometimes statist solutions that would be hostile to the economic model of the platform companies.

The possibility that platform power may be used to influence the political process, not only through lobbying but also by shaping public discourse, is especially concerning.

The four criteria outlined above provide a framework for evaluating platform accountability. As well as analysing accountability practices, researchers need to assess the quality of the relations of justification that they enact. Are they inclusive? Are they based on transparent information? Do they promote deliberative reflection? Do they empower affected publics? We can evaluate specific accountability practices using these criteria, identifying strengths as well as weaknesses. Since different practices may realize particular principles but not others, we can consider how practices might work together to enact better relations of justification overall, as systemic approaches to deliberative democracy advocate (Mansbridge et al., 2012). Finally, researchers may anticipate better relations of justification in their own research. Deliberative methods, such as focus groups or deliberative workshops, can carve out spaces for publics to reflect on platform governance in ways that existing accountability practices may not allow.

Conclusion

Fuelled by regular public controversies, the debate about platform power looks set to continue. Of course, researchers will have their own views about platform governance, but they cannot have the final word in this debate. Ultimately, answers to the question of how platforms are governed can only be arrived at democratically through a public discourse that involves all those who may be affected. But then researchers can point critically to shortcomings (exclusions, information deficits and limited opportunities for reflection and choice) in current public discourse. We can also reflect on how these short-comings may be addressed and genuine accountability achieved. The current concern about platform power in fact betrays a deeper political problem: how the 'justificatory power' (Forst, 2014b) of publics in relation to the governance of the digital can be enhanced. We hope that the research agenda we have outlined here might make a small contribution to wider efforts to address this problem.

Note

1.   Our criteria (inclusion, transparency, deliberation and empowerment) match four of the 'goods' Smith (2009) uses to evaluate participatory-democratic initiatives (inclusiveness, transparency, considered judgement and popular control), but they are tailored to and interpreted within the specific context of platform accountability. We do not have space to consider Smith's criteria of efficiency and transferability.

References

Ananny, M. and Gillespie, T. (2017) 'Public Platforms: Beyond the Cycle of Shocks and Exceptions', Oxford Internet Institute. Available from http://blogs.oii.ox.ac .uk/ipp-conference/sites/ipp/files/documents/anannyGillespie-publicPlatforms-oii -submittedSept8.pdf/ (accessed 24 September 2019).

Andrejevic, M. (2013) 'Public Service Media Utilities: Rethinking Search Engines and Social Networking as Public Goods', *Media International Australia Incorporating Culture and Policy*, 146(1), pp. 123–32.

Andrejevic, M. (2014) 'Big Data, Big Questions: The Big Data Divide', *International Journal of Communication*, 8, pp. 1673–89.

Baker, C.E. (2002) *Media, Markets, and Democracy*. Cambridge: Cambridge University Press.

Baker, C.E. (2007) *Media Concentration and Democracy: Why Ownership Matters.* Cambridge: Cambridge University Press.

Bardoel, J. and d'Haenens, L. (2004) 'Media Responsibility and Accountability: New Conceptualizations and Practices', *Communications*, 29(1), pp. 5–26.

Barwise, P and Watkins, L. (2018) 'The Evolution of Digital Dominance: How and Why We Go to GAFA'. In M. Moore and T. Tambini (eds), *Digital Dominance: The Power of Google, Amazon, Facebook, and Apple.* Oxford: Oxford University Press, pp. 21–49.

Bunting, M. (2018) 'From Editorial Obligation to Procedural Accountability: Policy Approaches to Online Content in the Era of Information Intermediaries', *Journal of Cyber Policy*, 3(2), pp. 165–86.

Crawford, K. and Gillespie, T. (2014) 'What Is a Flag for? Social Media Reporting Tools and the Vocabulary of Complaint', *New Media and Society*, 18, pp. 410–28.

DeNardis, L. and Hackl, A.M. (2016) 'Internet Control Points as LGBT Rights Mediation', *Information, Communication and Society*, 19(6), pp. 753–70.

Dencik, L. (2018) 'Surveillance Realism and the Politics of Imagination: Is There No Alternative?', *Krisis: Journal for Contemporary Philosophy*, 1, pp. 31–43.

van Dijck, J., Poell, T. and de Waal, M. (2018) *The Platform Society: Public Values in a Connective World.* Oxford: Oxford University Press.

Draper, N.A. and Turow, J. (2019) 'The Corporate Cultivation of Digital Resignation', *New Media and Society*, 21(8), pp. 1824–39.

Eberwein, T., Fengler, S. Lauk, E. and Leppik-Bork, T. (2011) *Mapping Media Accountability – in Europe and Beyond.* Köln: Herbert von Halem.

Flew, T., Martin, F. and Suzor, N. (2019) 'Internet Regulation as Media Policy: Rethinking the Question of Digital Communication Platform Governance', *Journal of Digital Media and Policy*, 10(1), pp. 33–50.

Forst, R. (2014a) *The Right to Justification: Elements of a Constructivist Theory of Justice.* New York: Columbia University Press.

Forst, R. (2014b) *Justification and Critique: Towards a Critical Theory of Politics.* Cambridge: Polity.

Forst, R. (2017) *Normativity and Power: Analyzing Social Orders of Justification.* Oxford: Oxford University Press.

Freedman, D. (2008) *The Politics of Media Policy.* Cambridge: Polity.

Freedman, D. (2016) The Internet of Rules: Critical Approaches to Online Regulation and Governance. In J. Curran, N. Fenton and D. Freedman (eds), *Misunderstanding the Internet.* London: Routledge, pp. 117–44.

Giddens, A. (1984) *The Constitution of Society: Outline of the Theory of Structuration.* Cambridge: Polity.

Gillespie, T. (2010) 'The Politics of "Platforms"', *New Media and Society*, 12(3), pp. 347–64.

Gillespie, T. (2018) *Custodians of the Internet: Platforms, Content Moderation, and the Hidden Decisions that Shape Social Media.* New Haven, CT: Yale University Press.

Habermas, J. (1997) *Between Facts and Norms: Contributions to a Discourse Theory of Law and Democracy.* London: Polity Press.

Hirschman, A. (1970) *Exit, Voice, and Loyalty: Response to Decline in Firms, Organizations, and States.* Cambridge, MA: Harvard University Press.

Lessig, L. (2006) *Code 2.0.* New York: Basic Books.

Mansbridge, J. Bohman, J., Chambers, S., Christiano, T., Fung, A., Parkinson, J., . . . Warren, M. (2012) A Systemic Approach to Deliberative Democracy. In J. Parkinson and J. Mansbridge (eds), *Deliberative Systems*. Cambridge: Cambridge University Press, pp. 1-26.

McQuail, D. (2005) *Media Accountability and the Freedom of Publication*. Oxford: Oxford University Press.

Moore, M. and Tambini, D. (2018) *Digital Dominance: The Power of Google, Amazon, Facebook, and Apple*. Oxford: Oxford University Press.

Moss, G. (2018) 'Media, Capabilities, and Justification', *Media, Culture and Society*, 40(1), pp. 94–109.

Smith, G. (2009) *Democratic Innovations: Designing Institutions for Citizen Participation*. Cambridge: Cambridge University Press.

Tambini, D. (2018) 'Social Media Power and Election Legitimacy'. In M. Moore and D. Tambini (eds), *Digital Dominance: The Power of Google, Amazon, Facebook, and Apple*. Oxford: Oxford University Press, pp. 265-93.

Thompson, M. (2016) 'Beyond Gatekeeping: The Normative Responsibility of Internet Intermediaries', *Vanderbilt Journal of Entertainment and Technology Law*, 18, p. 783.

Yeung, K. (2017) '"Hypernudge": Big Data as a Mode of Regulation by Design', *Information, Communication and Society*, 20(1), pp. 118–36.

10 Human rights futures and the digital: a radical research agenda

M.I. Franklin

Introduction

What do universities, freedom of expression in light of anti-hate speech measures, the School Strike for Climate, the #GeziPark and #MeToo movements, and 'fake news' polemics have in common? The answer may seem obvious: they are all beholden to the digital and globally networked environments in which participants seek to inform, persuade and mobilize others. This truism belies the intensity of the world's designed dependence on computer-dependent media and communications technologies, and institutionalized reliance on their increasingly automated transmission infrastructures enveloping the planet: private and government-owned submarine tubes, radio masts and fibre optic cables on land, and satellites orbiting the Earth. This co-dependence also affects approximately 3 billion of the world's 7.7 billion inhabitants, for whom a predetermined relationship between internet access, human rights and 'development' have become encapsulated in the United Nations' (UN) Sustainable Development Goals (UN, 2015).

This situation reaches back at least 50 years of primarily United States (US)-led research and development (R&D) into 'the digital', a genus of information and communications technologies (ICT). From the 1980s, the internet's packet-switching architecture and computing protocols were being rolled out, personal computers becoming a consumer item, national telecommunications operators privatizing and digitalizing, and World Wide Web applications opening up a means of connecting across national borders and localities to the general public; computer-mediated communications that were once the preserve of computer scientists and elite research institutions. Yet it is only since the second decade of this century that the international human rights community, the UN's human rights and humanitarian agencies, alongside non-governmental organizations such as Amnesty and Human Rights Watch, have committed to maintaining an online, networked presence. These organizations now recognize their obligation to educate staff about the implications that social media tools and internet-based sources of information, via smartphones in particular, have for their own rights and freedoms (for example, privacy), let alone those of their constituencies (for example, freedom of assembly, freedom of expression). They have become gradually more active in human rights advocacy for internet policy agenda-setting in 'multistakeholder internet governance' consultations; for example, the Internet Governance Forum and comparable agendas in multilateral institutions such as the World Economic Forum. With the setting-up of technology and human rights, and digital rights programmes (Amnesty and Human Rights Watch, respectively), these once digital sceptics acknowledge that human rights are as fragile, and need as much defending, online as offline (Elsayed-Ali, 2015; UNHRC, 2014; Council of Europe, 2014).

Rationale

This chapter presents the outline of a future research agenda, one that can more fully apprehend the theoretical and empirical dimensions to the (un)intentional entanglements of traditional human rights advocacy, ICT, R&D and concomitant 'digital agendas' at the national and intergovernmental level, 'digital rights' discourses and jurisprudence (Rosen, 2012; Bolton, 2015; Dutton et al., 2011). As (social) media panics about smartphone use for post-millennial generations of 'digital natives' make headlines, the need to advance a fuller spectrum of research into the societal and environmental impacts of these governmental and corporate investments in future, of emerging digital technology designs for cyber-security and commercial purposes, converges on gaps in the know-how and 'want-to' of existing human rights monitoring bodies, legislatures and judiciaries.

By 'future' I am referring to a theoretical agenda that addresses critical paradigms that have yet to take into account this techno-societal and human rights nexus when theorizing 'the digital', as cultures of internet use, digital or online content, and dynamics of oppression and exclusion. This means taking burgeoning scholarship at the intersection of these domains to meet the race, gender and class dimensions exposed by 'decolonizing' critiques of academe, and social and political institutions (Benhabib, 1997; Kaseem, 2019; Noble,

2018; Tuhiwai Smith 1999). These critiques are indispensable to apprehending how the link between documented 'digital divides' and technological R&D perpetuate race, class, gender and community-based structures of discrimination, exclusions that continue to be inadequately addressed within internet governance consultations that link digital policy-making with human rights jurisprudence. A third aspect to this pluriform disciplinary agenda is to take on board cutting-edge theories and research into the legal ramifications ('soft' and 'hard' law) of digital age phenomena; for example, privacy online, cyberspace, jurisdiction, the 'right to be forgotten', and biometrics. This agenda thereby includes more robust *constructivist* understandings of the interplay between the transformative legacy of digital and networked technologies to date, socio-political agency, and incumbent techno-economic power (Feenberg, 1999; Appadurai, 2002). This strategic gathering of intellectual and advocacy forces from diverse traditions can then engage with theories of radical democracy. These theories address post-Cold War and War on Terror limitations on what is still possible for participatory and inclusive democratic praxis, and do so without falling into the traps of techno-economic fatalism, fetishized individualism or universalizing essentialisms (Laclau and Mouffe, 2001; Dahlberg and Siapiera, 2007; Mouffe, 2013).

Such an agenda can then engage with the legal detail and practical considerations of the shifting form, and substance of existing human rights law and norms, as technological experiments – marketed as inevitable innovations – gallop on ahead of courts, regulators and much of the general populace. The second, empirical linchpin is to propose and execute unapologetically multidisciplinary and multi-sited methodologies that can address these questions in order to withstand the undertow of disciplinary silos and conceptual cul-de-sacs. As others have been arguing, for at least as long as R&D into digital technologies and consumer products (cybernetics, robotics, biotech and household appliances to name a few) has gone hand in hand with the internet timeline, hi-tech societies already have 'cyborg', if not 'post-human' characteristics through these human–machine entanglements (Haraway, 1990; Hayles, 1999; Holmes, 2007). This agenda does not shy away from considering the changing status of 'being human' in current international human rights law as 'we' – sentient creatures of flesh and blood, claiming autonomy, consciousness and rational thought – soporifically consent to being reconfigured as bots, surveillance data and social media profiles. These compacts, which we make with our digital selves and service providers, have already been enrolled into a range of intra- and extra-corporeal systems of everyday cyberveillance based on personal biometric data-harvesting, ethnic profiling algorithms and behavioural modification (Kulesza and Balleste, 2015).

Such observations need not reaffirm the doomsday scenarios popular in dystopian science fiction, film and television, and scary book titles. But they do beg the question of whether the enforcement of current legal norms and standards is sufficient for research and policy agendas addressing the complex, co-constitutive dynamics between 'the digital' and existing human rights as rule of law. Debates about whether internet access should be construed as a (new) right, according to which legal criteria or precedents, is one example (Jørgensen, 2013; Franklin, 2013). Another case in point is the debate around the legal status of cyberspace or 'internet jurisdiction' within an international·system and rule of law premised on the territorial sovereignty of the Westphalian nation-state (Tsagourias, 2015; Fraser, 2007; Franklin, 2018, 2019).

Problematic

At this historical juncture, this research agenda acknowledges a tripartite tension. First, it recognizes that human rights law-making and jurisprudence are, by definition, cautious enterprises, so they are conservative in temporal terms. Any 'new' rights emerge slowly, their status as customary or hard law coloured by the respective levels of official endorsement. The 1949 Universal Declaration of Human Rights is one case in point. Furthermore, the world's planetary monopolies in this sector, 'tech giants' based largely in the US but with strong competition from state-owned versions in the People's Republic of China and Russia, are teaming up with local and national legislatures, in the Global South particularly, looking for the next silver bullet to solve any manner of socio-economic and service provision inadequacies. These actors move a lot more quickly than do judiciaries and intergovernmental organizations, their motto being, in the words of Mark Zuckerberg, chief executive officer of the global corporation called Facebook, 'move fast and break things'. Sandwiched between these forces of order, ownership and control are ordinary people, users and non-users of digital, networked computing, mobile devices and self-regarding commercial social networks.

Second, it recognizes that as legal standards and ideals, international human rights treaties and covenants are coterminous with the rise of Western liberal models of democracy in the second half of the twentieth century. Always political, and easily politicized, twentieth-century human rights discourses and the laws that underpin them as international norms remain contested and unevenly ratified (Vincent, 2010). This is because states, as (non-)signatories to UN treaties and covenants, also perpetrate human rights abuses. Furthermore, since the 1980s, telecommunications operations and computing technology development have been devolved to private actors, particularly in parts of the world where telecommunications used to be public services. The global corporations that own and control the lion's share of the world's internet access and mobile devices are not signatories to these international treaties.

Third, traditional human rights advocacy agendas constitute a worldview, one in which universal rights are conceived as trans-cultural and trans-historical phenomena, encapsulated through literal readings of human rights law. A comparable worldview undergirds the policy discourses constituting 'Global Internet Governance' consultations, substituting 'hard law' with principles of 'global consensus-building' and inclusive, rather than expert participatory models. Despite their respective lexicons of certainty and historical destiny, these domains remain contested ground.

Plotting an alternative course

Five foundational observations are integral to any research focus within this agenda's remit for future inquiries, given the above tensions. Human rights, conventionally defined within the UN's International Bill of Rights and subsequent treaties and covenants, entail:

1. A complex and diverse body of customary international law, legal instruments and standards, with varying degrees of endorsement from UN member-states.
2. An advocacy framework based on the notion of a 'global civil society', non-governmental organizations and other organizations concerned with what are either primarily Anglo-American civil liberties discourses, or the fuller spectrum of human rights law over three to four generations of international treaties and covenants.
3. A foundational narrative underpinning all UN undertakings, particularly since the 2000 Millennium Goals (UN, 2000). This narrative has now become part of a digitally networked paradigm for global development under the 2015 Sustainable Development Goals (UN, 2015; UNHRC, 2014).
4. A domain of intensifying techno-economic and transnational contestation around the ownership and control (regulation) of future digital, networked media and communications technologies; transmission design, terms of access and use, content regulation, (personal and group) data storage and management.

5. Emerging issues in the wake of the 2015 Paris Agreement on Climate Change, from which the US is due to withdraw in November 2020; and transnational mobilization of young people around the world, for example, the School Strike for Climate and Existence Rebellion movements. These events throw into sharp relief the co-dependence between internet use and non-renewable energy sources, environmentally degrading materials, and inhumane manufacturing practices.

These dimensions point to how interpreting, let alone enforcing existing human rights law for the online environment, now and in the future, cannot be achieved in isolation. Local, national and international policy-makers have been forging public–private agreements with today's tech giants to refine excessively intrusive surveillance tools and systems for public spaces, alongside specific national and cyber-security applications based on a commitment to the next generation of artificial intelligence (AI), the commercial and civil roll-out of an 'Internet of Things', 'smart' and 'digital city' action plans. All these undertakings presuppose that the daily running of the internet – its strategic infrastructure, service delivery, human–machine and human-to-human interactions – can be devolved to automated devices and systems without affecting even existing rights and freedoms; an error of judgement based on mutually reinforcing ignorance and failures in foresight.

Paradigm shifting: future visions

Chantal Mouffe's (2013) conceptualization of 'agonistics' provides a substantive perspectival shift for moving this agenda forward. This radical democracy project looks to recuperate democratic ideals and working principles from the bloodshed of both Cold War and War on Terror geopolitics. It is based on recognizing that the world is 'a pluri-verse not a universe' (ibid.: 22), and the rejection of the dogmatic 'implementation of the Western liberal democratic model' (ibid.: 29) at the expense of all others. It argues for an organizational – and conceptual – ethos that construes any democratic project as a 'form of an "agonism" (struggle between adversaries)' rather than an inevitable cycle of '"antagonism" (struggle between enemies)' (Mouffe, 2013: 7). Claiming, as does much of 'multistakeholder internet governance' rhetoric, that outcomes based on 'rough consensus' imply the full consent and participation of 'everyone' belies the accompanying erasure of substantive dissenting views from the official record. This is tantamount to negating the 'possibility of legitimate dissent, thereby creating a favourable terrain for the emergence of violent forms of antagonism' (ibid.: 20). As Mouffe argues, a democratic digital politics:

> is not and cannot be the domain of the unconditional because it requires making decisions in an undecidable terrain. This is why the type of order which is established through a given hegemonic configuration of power is always a political, contestable one; it should never be justified as dictated by a higher order and presented as the only legitimate one (ibid.: 17).

The object of this trenchant, provocative critique of liberal democratic virtue is the 'naïve' cosmopolitanism underpinning an understanding of 'consensus [that] presupposes the existence of a political community which is simply not available at the global level. Indeed, to envisage the world order in terms of a plurality of hegemonic blocks requires relinquishing the idea that they need to be part of an encompassing moral and political unit' (ibid: 23). This envisioning of 'agonistic pluralism' includes the recognition that within existing human rights treaties there is room for 'a pluralism of interpretations' (ibid: 30).

The implication of Mouffe's formulations for this heterodox research agenda is that we live in not only a 'multipolar agonistic world', but also one constituted by multiplex digitalized and/or networked encounters, spaces and power distributions. In this sense, the promises and limits of existing human rights legalities, as these are now playing out as potentially 'new' rights for the online–offline nexus, need to be formulated 'in a way that permits a pluralism of interpretations [and] . . . recognition of a plurality of forms of democracy [to ensure] a just social and political order . . . within and across diverse cultures and polities' (ibid.: 30–31).

A radical research agenda

Digital and networked (cyber)spaces have become officially recognized as co-constituents of everyday life, politics and business, albeit in ways that exceed Westphalian conventions of 'prescription, enforcement, and adjudication' for traditional, sovereign powers (Tsagourias, 2015: 20). There are two further insights to incorporate into the conceptual framework of this agenda as a consequence. First, the way that people use the internet and, conversely, the ways in which competing actors deploy these designs for their respective agendas in turn (civic, commercial and criminal) are rewiring the rule of law and understandings of democratic oversight that underpin state sovereignty as a geographical, territorial entitlement, and jurisdiction over citizens and residents. Customary territorial sovereignty, as the 'basic organizing principle of international law' (Tsagourias, 2015: 15) has already given way to how digitalized 'networks of networks' governed by intricate layers upon layers of proprietary computer codes – software designs – arguably constitute de facto 'laws' of another order (Lessig, 2006; Marsden, 2011).

Second, if humans, construed in liberal theory and governmental practice as autonomous individuals, are still taken as the subject of international human rights law, then who or what are to be the subjects of would-be digital rights? One could argue that bots, robots, androids and other forms of emerging artificial intelligences, on-screen avatars and personae, designed to mimic (respond to or as) human agents have rights as well. This is not the stuff of science fiction. These are real-world ethical and legal issues that artists, novelists and philosophers have been more willing to consider than have engineers, policy-makers, legal scholars and social scientists. So, the next step is to develop research that can break open any one of, or combination of, the following conceptual black boxes:

1. 'Governance', particularly when it is formulated for so-called internet governance, tends to be treated as a neutral descriptor for a self-explanatory policy, and disciplinary field. This is not the case, considering the deeper rifts undergirding consultations that require state actors and corporate powers to loosen their grip on the reins of either legislative or economic power.
2. 'Technology' is also another term that needs reconsideration, unpacking and conceptualizing as socially and historically constituted, even when the societal effects, and the inevitability of technological roll-out appear to transcend social agency or political will. Philosophical debates about the substantive nature of major technological change (for example, the printing press, railways, electricity and mechanization) are not given sufficient attention in research agendas addressing the societal and political impact of digital and networked technologies as endogenous rather than exogenous phenomena.
3. 'Rights', whether these are construed as civil liberties, selected 'fundamental rights and freedoms' (Council of Europe, 2014) or international human rights standards, are also narratives and powerful framing devices that, even when written out and enforced as law, are polysemic in interpretation and affect. Human rights may be, literally speaking, universal but this does not make them absolute in practice.
4. 'Participation' as a proxy for consultative democracy has been a focus of research in policy consultations based on 'multistakeholder participatory

models'. Even when tempered by diverging access to resources, know-how and other entitlements that come with being able to get to the table, this working principle needs closer attention rather than dismissal as doomed to failure.

From here a specific focus – for the topics and issues are myriad in this domain – needs to develop a methodology along the following axes:

1. Temporal – longer-term and fuller immersion in the field.
2. Spatial – multi-sited points of access and presence, data-gathering and analysis at the online–offline nexus.
3. Practice – attention paid to processes matters as well as static forms of 'content' or 'discourse' analysis of official outcomes, outputs and transcripts; staying close and/or panning out to appreciate institutional formations imply another scale of practice.
4. Techno-economic indicators – ownership matters in heavily patrolled copyrighted domains.
5. Geocultural sensibilities – stepping outside the Western orbit of theory and research, policy-speak and commercial interests reveals viable and alternative communities and cultures of use, thereby opening up how human and digital rights are experienced as given, or as aspirations.
6. Environmental factors – digitally induced, sociologically conceived and physically implicated, as all these dimensions are interconnected by and through computing networks and their unsustainable energy requirements.

Successive revelations of extra-legal, state-sanctioned and corporate-owned routines of digitalized and online surveillance and data-gathering have been a game-changer. These practices corrode the foundations upon which the rights and freedoms that Western polities consider as indispensable to any well-functioning democratic society rest. First, this is observable in the shifting terms of reference and domains in which 'rights', however defined, in digitally networked contexts, are being contested: enjoyed or undermined. Second is the fact that conventional notions of territorial sovereignty that UN member-states presume they can claim over 'cyberspace' and related digital domains pale in comparison to the reach and depth of ownership and control that incumbent 'commercial sovereigns' already exercise over everyday life, business and politics at the online–offline nexus.

Conclusion

High-level proclamations about human rights principles across digital policy-making arenas actually throw the fragility of existing human rights standards, online and on the ground, more into relief. This research agenda takes into account this double standard between lip service paid to, and substantive advances in, 'human rights by design' outcomes. Doing so means getting past the epistemological habits and political limitations of reliance on comfortable dichotomies; for example, human versus machine, virtual versus real, online versus offline. This is a challenge, given the entrenched positions between proponents of 'internet regulation' on the one hand and, on the other, market forces. It is also a challenge in the face of long-standing cynicism about the practical realities of multi-interdisciplinary research that is also socio-politically engaged. Nonetheless, for the next generation of ICT roll-out and quality of life for tomorrow's digital natives who reside within and outside the internet heartlands of the world's richest economies, such an agenda can break new ground. It can contribute to mobilizations challenging the 'no alternative' motto in which contemporary manifestations of digital networks become forms of 'rational imperialism [that prove] to be a façade for cynical imperialism' (Koskenniemi, cited in Mouffe, 2013: 36). And as signs of 'digital imperialism' come to light, in terms of global monopolies and state intrusions taking place on a scale not yet seen, this agenda can open up avenues for investigating democratically viable and environmentally sustainable futures for future generations.

References

Appadurai, A. (2002) 'Disjuncture and Difference in the Global Cultural Economy'. In J.X. Inda and R. Rosaldo (eds), *The Anthropology of Globalisation: A Reader*. Malden, MA, USA and Oxford, UK: Blackwell, pp. 27–47.

Benhabib, S. (1997) 'Strange Multiplicities: The Politics of Identity and Difference in a Global Context', *Macalester International*, 4, pp. 27–56.

Bolton, R.L. (2015) 'The Right to Be Forgotten: Forced Amnesia in a Technological Age', *John Marshall Journal of Information Technology and Privacy Law*, 133, pp. 133–44.

Council of Europe (2014) *The Rule of Law on the Internet and in the Wider Digital World*, December. Strasbourg: Council of Europe Commissioner for Human Rights. Available at https://wcd.coe.int/ViewDoc.jsp?Ref=CommDH/IssuePaper%282014%291&Language=lanEnglish&Ver=original&Site=COE&BackColorInternet=DBDCF2&BackColorIntranet=FDC864&BackColorLogged=FDC864/ (accessed 26 August 2019).

Dahlberg, L. and Siapiera, E. (eds) (2007) *Radical Democracy and the Internet: Interrogating Theory and Practice*. London, UK and New York, USA: Palgrave Macmillan.

Dutton, W.H., Dopatka, A., Hills, M., Law, G. and Nash, V. (2011) *Freedom of Connection – Freedom of Expression: The Changing Legal and Regulatory Ecology Shaping the Internet*. Paris: UNESCO.

Elsayed-Ali, S. (2015) 'Amnesty International Responds To UK Government Surveillance', *Intercept*, 2 July. Available at https://firstlook.org/theintercept/2015/07/02/op-ed-amnesty-international-responds-u-k-government-surveillance/ (accessed 26 August 2019).

Feenberg, A. (1999) *Questioning Technology*. New York, USA and London, UK: Routledge.

Franklin, M.I. (2013) *Digital Dilemmas: Power, Resistance and the Internet*. London, UK and New York, USA: Oxford University Press.

Franklin, M.I. (2018) 'Refugees and Digital Gatekeepers of 'Fortress Europe', State Crime and Digital Resistance: Special Issue of the International State Crime initiative, Anne Alexander and Saeb Kasm (eds), *State Crime Journal*, 7(1), pp. 77–99.

Franklin, M.I. (2019) 'Human Rights Futures for the Internet' In B. Wagner, M. C. Kettemann and K. Vieth, eds, *Research Handbook on Human Rights and Digital Technology*. Cheltenham, UK and Northampton, MA, USA: Edward Elgar Publishing, pp. 5–23.

Fraser, N. (2007) 'Transnationalizing the Public Sphere: On the Legitimacy and Efficacy of Public Opinion in a Post-Westphalian World', *Theory, Culture, and Society*, 24(4), pp. 7–30.

Haraway, D.J. (1990) 'A Manifesto for Cyborgs: Science, Technology, and Socialist Feminism in the 1980s'. In L. Nicholson (ed.), *Feminism/Postmodernism*. London: Routledge, pp. 190–233.

Hayles, K. (1999) *How We Became Posthuman: Virtual Bodies in Cybernetics, Literature and Informatics*. Chicago, IL: University of Chicago Press.

Holmes, B. (2007) 'Future Map or How the Cyborgs Learned to Stop Worrying and Learned to Love Surveillance', *Continental Drift: The Other Side of Neoliberal Globalization* (blog). Available at http://brianholmes.wordpress.com/2007/09/09/future-map/ (accessed 26 August 2019).

Jørgensen, R.F. (2013) *Framing the Net: The Internet and Human Rights*. Cheltenham, UK and Northampton, MA, USA: Edward Elgar Publishing.

Kaseem, A. (2019) 'Marching towards Decolonisation: Notes and Reflections', *Sociological Review*, 28 July. Available at https://www.thesociologicalreview.com/marching-towards-decolonisation-notes-and-reflections/.

Kulesza, J. and Balleste, R. (eds) (2015) *Cybersecurity: Human Rights in the Age of Cyberveillance*. Lanham, MD: Rowman & Littlefield/Scarecrow Press.

Laclau, E. and Mouffe, C. (2001) *Hegemony and Socialist Strategy: Towards a Radical Politics*. London, UK and New York, USA: Verso.

Lessig, L. (2006) *Code and Other Laws of Cyberspace, Version 2.0*. New York: Basic Books.

Marsden, C.T. (2011) *Internet Co-Regulation: European Law, Regulatory Governance and Legitimacy in Cyberspace*. Cambridge: Cambridge University Press.

Mouffe, C. (2013) *Agonistics: Thinking the World Politically*. London, UK and New York, USA: Verso.

Noble, S.U. (2018) *Algorithms of Oppression: How Search Engines Reinforce Racism*. New York: New York University Press.

Rosen, J. (2012) 'The Right to Be Forgotten', *Stanford Law Review* 64(88), 13 February. Available at https://review.law.stanford.edu/wp-content/uploads/sites/3/2012/02/64 -SLRO-88.pdf (accessed 14 February 2020).

Tsagourias, N. (2015) 'The Legal Status of Cyberspace'. In N. Tsagourias and R. Buchan (eds), *Research Handbook on International Law and Cyberspace*. Cheltenham, UK and Northampton, MA, USA: Edward Elgar Publishing, pp. 13–29.

Tuhiwai Smith, L. (1999) *Decolonising Methodologies: Research and Indigenous Peoples*, London UK: Zed Books.

UN (2000) *Millennium Development Goals*. UN General Assembly. Available at http://www.un.org/millenniumgoals/ (accessed 26 August 2019).

UN (2015) *Sustainable Development Goals*. UN General Assembly. Available at http://www.un.org/sustainabledevelopment/sustainable-development-goals/ (accessed 26 August 2019).

UN Human Rights Council (UNHRC) (2014) *Resolution A/HRC/26/L.24: Promotion and Protection of All Human Rights, Civil, Political, Economic, Social and Cultural Rights, Including the Right to Development*, Twenty-sixth session, Agenda item 3, UN General Assembly, 20 June. Available at http://ap.ohchr.org/documents/dpage _e.aspx?si=A/HRC/26/L.24/ (accessed 26 August 2019).

Vincent, A. (2010) *The Politics of Human Rights*. Oxford: Oxford University Press.

PART IV

Informational, Symbolic and Communicative Actions

11 After clicktivism

Dave Karpf

In September 2010, *The New Yorker* published an article by Malcolm Gladwell titled 'Small Change', in which the author bemoaned the fate of activism in the age of the Internet: 'Where activists were once defined by their causes, they are now defined by their tools . . . We seem to have forgotten what activism is.' Gladwell's article captured the zeitgeist of the moment, in which researchers and journalists were wondering aloud about the dangers of 'clicktivism' (or 'slacktivism') and asking how the new digital activism compared to older forms of street activism. Is online engagement too easy? Does it increase or decrease the likelihood of taking more costly political actions? Is it effective? Are there downsides to the apparent democratization of political speech online? Clicktivism would go on to be a central question within the research community for roughly a decade.

Gladwell's article did not begin the clicktivism debate, but it became a centrepiece of the academic conversation for years to come. Despite being neither peer-reviewed nor scholarly, the article has been cited over 1500 times in the near-decade since, a testament to the gravitational pull of the clicktivism conversation. Clicktivism has been an empirical puzzle for some and a theoretical stalking horse for others. Over the years it has repeatedly been challenged, debunked, deconstructed and rethought. I myself have been a repeat contributor, with an article and two books arguing that we can only make sense of the efficacy of digital tactics if we pay attention to the organizational layer of political contention. It was a basic observation, but one that seemingly needed to be repeated, time and time again. Research conversations move slowly, and tend to become anchored in high-profile provocations like Gladwell's. Over the years, many of us have made similar points in parallel with one another, each taking a turn theoretically upending or empirically disproving some element of Gladwell's argument.

Mercifully, it appears to me that the academic debate over clicktivism has finally settled down. A decade was perhaps long enough. The field has come to adopt a more nuanced perspective on the subject, with authors such as James

Dennis (2019) and Jen Schradie (2019) profoundly demonstrating that the affordances of the social media for political activism cannot simply be mapped onto an axis of bad-to-good. It has certainly helped that the aftermath of Trump's election and the Brexit referendum have led the research community (with its long-standing habit of oversampling cases from the United States and the United Kingdom) to wonder about the rise of right-wing populism and the explosion of mass political street protests. The clicktivism concern now feels dated, because it is hard to claim in the midst of record-setting political demonstrations that citizens are too content to 'like', tweet, and sign e-petitions.

The question I ponder in this chapter, then, is: what comes next? Where might the attention of the research community productively turn if we are finally ready to move beyond clicktivism and slacktivism? This chapter proposes four areas that I believe are ripe for further research. First, researchers have begun to collaborate productively with advocacy groups themselves, building the same types of partnerships that have flourished for several years with electoral campaigns (Issenberg, 2012). This type of collaboration has tremendous potential for advancing public knowledge while also providing value to worthwhile non-governmental organization (NGO) initiatives. Second, researchers have the opportunity to take a more holistic approach to the study of digital activist networks, tracing the resource flows between organizations, media platforms, funders, political parties and technologists to map the landscape of contemporary activism. Digital trace data can enhance our capacity to visualize these activist networks. Third, researchers have begun to study the populist backlash among right-wing activist networks. Such research promises to counterbalance the long-standing habit of digital activism researchers to study left-wing movements. Fourth, researchers are just beginning to adopt a cross-national comparative framework, examining networked activist movements that run similar campaigns with similar tactics across different political cultures and electoral systems.

I then conclude by discussing the potential for researchers to ask the deeper, more complicated questions that past generations of researchers have largely avoided (Walker, 1991). In particular, the current moment calls for a reassessment of the very nature of political power. Clearly, defining power can be a quixotic endeavour, especially when spanning political systems, ideological orientations and activist tactical repertoires. But it is a worthy challenge, a better use of our collective attention than the past decade spent repeatedly overturning the clicktivism canard. I detail each of these potential areas of analysis below.

Embedded digital activism research

What often sets the study of digital activism apart from the broader field of social movement studies is the availability of publicly accessible digital trace data that can be used to analyse political participation. E-petitions, retweets, hyperlinks and Facebook 'likes' all provide new forms of networked data that can render political participation visible in new and interesting ways (Bennett and Segerberg, 2013; Margetts et al., 2016). While this remains a promising avenue for large-scale analysis of political behaviour and networked activism, it also has limitations that can be usefully augmented through embedded research in the years to come.

The problem with relying too much on digital trace data is the hidden biases it can introduce into our research. The most obvious of these biases is the habitual scholarly focus on Twitter over other, more heavily used social media sites. We know that Facebook, Google and YouTube are more popular than Twitter. We know that email is universally used (at least in the Global North countries where Twitter is most popular). But Twitter's application programming interface (API) is much easier to work with than Facebook's or YouTube's. Even in the age of 'big data', researchers are data-beggars rather than data-choosers. We construct research projects around the data we have access to, not the data we wish we had access to. Beyond the bias in favour of Twitter and hyperlink analyses, the research community's reliance on publicly accessible data creates key limitations because they force us to treat the algorithms of the major digital platforms as black boxes. Data scientists at Facebook, Google and Twitter are able to conduct far richer studies than traditional academic researchers, simply because they have access to proprietary data and can fashion studies that test the impact of tweaking the underlying algorithms that shape attention dynamics in the digital age (Karpf, 2019). Though a few researchers have begun to collaborate with these companies (Kramer et al., 2014; Bond et al., 2012), those collaborations are rare and can pose thorny ethical dilemmas.

An alternative route has been pioneered in recent years by scholars such as Hahrie Han and Adam S. Levine. Rather than analysing big, public data, these researchers are forging partnerships with civil society organizations and activist groups, thus getting access to the rich proprietary data that would otherwise be outside of our analytic field of vision. Han's (2014, 2016) work has broken new ground in analysing the different roles that mobilizing and organizing play in the power-building strategies of political associations. Levine and Kline (2018) conducted field experiments in partnership with social movement organizations, providing valuable insights into the types of frames that motivate supporters to get involved with the climate movement.

This trend of embedded research has roots in the study of American elections. Beginning with Donald Green and Alan Gerber's landmark voter turnout study (Green and Gerber, 2000), political scientists have increasingly embraced research designs that partner with electoral campaigns to run sophisticated field experiments that benefit both the electoral campaign and the broader research field. The Analyst Institute has been a particularly important forum for supporting these research partnerships (Issenberg, 2012). Both Han and Levine have founded new institutions that apply the Analyst Institute's model of embedded research to the study of social movements and NGOs. Han is a co-founder of the P3 research lab and is the director of the Agora Institute at Johns Hopkins University in the United States (US). Levine is the co-founder of Research for Impact (r4impact.org), which provides a 'matchmaking' service between researchers and advocacy organizations to facilitate embedded research.

Neither Han nor Levine primarily focus on digital politics research. They are political scientists who study contemporary activism and political participation. They have, however, become contributors to the study of digital activism simply by virtue of trends in contemporary activism: practically all large-scale activism has at least some digital component to it. In the years to come, there is tremendous potential for digital activism researchers to build upon their model of embedded research, partnering with digital activist groups and social movement networks to answer from the inside questions that would be rendered invisible through external analysis.

Digital activism and field theory: tracking resource flows

Embedded digital activism research emphasizes collaboration with a single political association, generating public knowledge that would otherwise be left opaque. I believe there is also great potential in applying field theory (McAdam and Fligstein, 2012) to the study of digital activism. Field theory provides a template for studying the institutional structures, norms and resource flows that support and maintain the organizational layer of political action. How do professional activist leaders learn their craft? What are their typical career paths, and where are they recruited from? Which organizations constitute the centre of any given issue area? What are their theories of change? Who funds them? How and when do they work together? The clicktivism debate anchored our research community for a decade in an atomistic approach to activism that was tactically focused, largely ignoring the enduring structures that shape activist networks over the long term. Tracing resource flows – funding, technology and expertise – can help us to build a descriptive picture of how activist networks operate, learn, grow and/or decline.

Kreiss and Saffer (2017) provide a useful template for this type of research. Their study of innovation in US electoral campaigns examines the professional biographies of campaign staffers across multiple campaign cycles. This approach allows them to examine where staff are recruited from, how they build their expertise, and how they build a broader ecosystem of consultancies and training organizations. It reveals biographical differences between Republican and Democratic political operatives, highlighting biases that become ingrained in the broader party apparatus over time. In a follow-up study, Kreiss and Saffer (2019) note that Democratic electoral campaigns draw much more heavily from the ranks of elite, Ivy League universities than Republican electoral campaigns. This finding runs counter to many researchers' expectations, in turn opening a theoretically generative line of further inquiry.

Similar techniques could be applied to the study of activist networks over time. Where did the core activists from Occupy Wall Street come from, and what have they gone on to do? What relations are there between the #BlackLivesMatter activist network, the #MeToo movement and the gun reform movement? Do they attend similar trainings, rely on similar technologies or appeal to the same donor bases? What are the similarities and differences between these activist networks and those that preceded them?

The challenge in applying field theory to digital activism research is similar to the challenge of embedded research: compiling the relevant data is a daunting task. Electoral campaigns are required to report their funding sources and their organizational charts. Social movement networks have no such repository. Just analysing where major US philanthropic foundations invest their money is a complicated problem. Determining the boundaries of leadership teams within networked activist communities and tracing their membership over time is exceptionally hard work. But the potential is equally immense. If we can assemble digital trace data to help build these datasets over time, we will establish new avenues of research and theoretical puzzles that are derived from the empirical realities of contemporary activism.

Studying the resurgence of right-wing activism

It is fair to say that the bulk of digital activism research has often focused on left-wing social movements. What is less clear is whether this has been due to personal progressive political bias among researchers or to an innate tendency to chase headlines. I suspect it is the latter: in the aftermath of the Arab Spring and Occupy Wall Street, researchers flocked to study the role of digital technologies in these political movements. Researchers have also been drawn to the study of gun reform activists, #BlackLivesMatter and #MeToo. These all happen to be progressive movements, but they are also landmark political moments. Researchers likewise have devoted attention to the Tea Party movement and the Cambridge Analytica scandal. When conservative movements dominate public debate, conservative movements become objects of fascination for the research community.

The resurgence of right-wing activism in recent years is producing increased demand for high-quality research on this topic. The research community is beginning to grapple with the use of social media by anti-democratic movements. Nate Persily (2017) has asked whether democracy can 'survive' the Internet. Lance Bennett and Steven Livingston (2018) have warned that digital disinformation can disrupt democratic norms. Alexander Hertel-Fernandez (2019) has studied the networks of Koch-funded conservative activists that have conducted a grassroots takeover of local governments throughout much of the US. Jen Schradie (2019) argues that digital technology inherently favours the tactics and techniques that are most often deployed by conservative activists.

Between conservative 'meme warriors', the rise of the 'alt-right' and the spread of white nationalism in the US and around the globe, the rise of conservative online activism has become an object of public fascination. I suspect that this will be an area of significant research activity for years to come, and that it will lead the research community to produce a more complex picture of the differences and similarities between left-wing and right-wing activist networks.

Digital activism in cross-national comparison

There is also an opportunity to add a robust comparative dimension to the study of digital activism. The 'netroots' model of digital activism that I described in my 2012 book (Karpf, 2012) has now been replicated and modified in several countries, including Germany (Campact), Australia (GetUp!), the United Kingdom (38 Degrees), Canada (Leadnow), New Zealand (Action Station), Ireland (Uplift) and Sweden (Skiftet). These organizations partner with each other through a global network called the Online Progressive Engagement Network (OPEN). They exchange ideas, discuss strategies and tactics, collaborate on shared technological platforms and partner with each other on international campaigns. They train their staff with the same materials, and agree to an overarching set of core operating principles (Karpf, 2013). Nina Hall (2019a, 2019b) has launched an ambitious research agenda exploring how these digital NGOs leverage power through digitally enabled transnational advocacy networks. At a 2018 academic workshop organized by James Dennis and Nina Hall, over a dozen researchers met to discuss global trends in digital activism. I believe we are just now seeing the first tentative steps toward building a framework for assessing the effects of digital activism across national boundaries.

A key distinction here is that cross-national comparison provides a different analytic framework than studies of international or global activism. These are not global NGOs or social movement networks targeting global corporations in their campaigns. The member organizations of the OPEN network are all nationally based, applying the same rapid-response digital mobilization model within different countries with distinct cultures and electoral systems. They thus provide us with an outstanding opportunity to assess similar cases across national contexts. How does the netroots model change in parliamentary systems? How do similar organizations morph and adapt to countries with different legal regimes, different philanthropic and civic traditions, and different pressure points for citizen political action? By tracing the similarities and differences among these nationally based digital activist groups, we can extend the boundaries of knowledge radically beyond individual studies of a single digitally enabled organization or social movement. Now that the research community seems to have moved beyond the clicktivism debate and has begun taking the organizational layer of political contention seriously, we can seize opportunities to develop hypotheses that can be tested across national contexts.

Conclusion: the fundamental question of power

The most frustrating thing about the clicktivism debate is that it is divorced from any theory of power. If we begin with the question, 'Is (digital petitioning/hashtag activism/online boycotts/any other online form of political expression) effective?', then the answer will invariably be, 'Well, it depends.' It depends on what the activists are trying to achieve, who they are targeting, and how a single 'clicktivist' tactic fits into the broader sequence of tactics that make up an activist campaign. The iconic offline tactics of previous eras (marches and mass demonstrations) are no more or no less inherently effective than the digital tactics that make up the clicktivist toolset. A hashtag can be powerful or pointless; a march can just as easily be powerful or pointless.

Theories of power are complex endeavours. More than 60 years ago, Robert Dahl wrote that power is relational, contextual and resource-dependent (Dahl, 1957). Several other theorists have expanded further on this concept (Lukes, 1974; Gaventa, 1980). There is no settled definition, nor is there a simple empirical framework for assessing the power of a given tactic or a specific campaign in the digital age. I do not expect that the digital politics research community will radically improve upon the deep-rooted methodological problems with measuring power in advocacy and activism. But, at the very least, we ought to try.

What types of political activism benefit most from the affordances of social media? How is power being deployed and diverted? How do activist networks learn and adapt new tactics and strategies, and how do powerful institutions react and respond over time? Is there, as Jen Schradie provocatively argues, a conservative bias to digital media? And how have activist groups evolved alongside the changing digital media environment? There is still so much that the research community does not know about digitally mediated contentious politics. Some of it we may never know. But there is much we can learn, if we collectively start asking better questions.

The digital politics research community has spent much of the past decade debating clicktivism. There was little to debate there, but Gladwell's essay provided an easy starting point. Many of us arrived at the same conclusions, but took our own routes to get there. We can now lay the clicktivism puzzle to rest. This chapter has offered a few thoughts on the questions we might pursue next. There are a lot of them. There is still so much left to learn.

References

Bennett, W.L. and Livingston, S. (2018) 'The Disinformation Order: Disruptive Communication and the Decline of Democratic Institutions', *European Journal of Communication*, 33(2), pp. 122–39.

Bennett, W.L. and Segerberg, A. (2013) *The Logic of Connective Action: Digital Media and the Personalization of Contentious Politics*. New York: Cambridge University Press.

Bond, R.M., Fariss, C.J., Jones, J.J., Kramer, A.D.I., Marlow, C., Settle, J.E. and Fowler, J.H. (2012) 'A 61-Million-Person Experiment in Social Influence and Political Mobilization', *Nature*, 489, pp. 295–98. doi:10.1038/nature11421.

Dahl, R.A. (1957) 'The Concept of Power', *Behavioral Science*, 2(3), pp. 201–15. doi:10.1002/bs.3830020303.

Dennis, J. (2019) *Beyond Slacktivism: Political Participation on Social Media*. New York: Palgrave Macmillan.

Gaventa, J. (1980) *Power and Powerlessness: Quiescence and Rebellion in the Appalachian Valley*. Chicago, IL: University of Illinois Press.

Gladwell, M. (2010) 'Small Change: Why the Revolution Will Not Be Tweeted', *New Yorker*, 4 October. Available at http://www.newyorker.com/reporting/2010/10/04/101004fa_fact_gladwell?currentPage=all/ (accessed 10 November 2019).

Green, D. and Gerber, A. (2000) 'The Effects of Canvassing, Telephone Calls, and Direct Mail on Voter Turnout: A Field Experiment', *American Political Science Review*, 94(3), pp. 653–63.

Hall, N. (2019a) 'Norm Contestation in the Digital Era: Campaigning for Refugees Rights', *International Affairs*, 95(3), pp. 575–95.

Hall, N. (2019b) 'When do Refugees Matter? The Importance of Issue-Salience for Digital Advocacy Organizations', *Interest Groups and Advocacy*, 8(3), pp. 333–55.

Han, H. (2014) *How Organizations Develop Activists: Civic Associations and Leadership in the 21st Century*. New York: Oxford University Press.

Han, H. (2016) 'The Organizational Roots of Political Activism: Field Experiments on Creating a Relational Context', *American Political Science Review*, 110(2), pp. 296–307.

Hertel-Fernandez, A. (2019) *State Capture: How Conservative Activists, Big Businesses, and Wealthy Donors Reshaped the American States – and the Nation*. New York: Oxford University Press.

Issenberg, S. (2012) *The Victory Lab: The Secret Science of Winning Campaigns*. New York: Crown Books.

Karpf, D. (2012) *The MoveOn Effect: The Unexpected Transformation of American Political Advocacy*. New York: Oxford University Press.

Karpf, D. (2013) 'Netroots Goes Global', *Nation*, 4 November. Available at http://www.thenation.com/article/176700/netroots-goes-global/ (accessed 10 November 2019).

Karpf, D. (2019) 'The Internet and Engaged Citizenship', White Paper, American Academy of Arts and Sciences. https://www.amacad.org/publication/internet-and-engaged-citizenship/ (accessed 10 November 2019).

Kramer, A.D.I., Guillory, J.E. and Hancock, J.T. (2014) 'Experimental Evidence of Massive-Scale Emotional Contagion Through Social Networks', *PNAS*, 111(24), 8788–90. doi:10.1073/pnas.1320040111.

Kreiss, D. and Saffer, A.J. (2017) 'Networks and Innovation in the Production of Communication: Explaining Innovations in US Electoral Campaigning from 2004 to 2012', *Journal of Communication*, 67(4), pp. 521–44.

Kreiss, D. and Saffer, A.J. (2019) 'Ivy League Democrats and State School Republicans', *Medium.com*. Available at https://medium.com/@dkreiss/ivy-league-democrats-and-state-school-republicans-8941960686d2/ (accessed 10 November 2019).

Levine, A.S. and Kline, R. (2018) 'Loss-Framed Arguments Can Stifle Political Activism', *Journal of Experimental Political Science*, 6(3), pp. 171–9.

Lukes, S. (1974) *Power: A Radical View*. New York: NYU Press.

Margetts, H., John, P., Hale, S. and Yasseri, T. (2016) *Political Turbulence: How Social Media Shape Collective Action*. Princeton, NJ: Princeton University Press.

McAdam, D. and Fligstein, N. (2012). *A Theory of Fields*. New York: Oxford University Press.

Persily, N. (2017) 'Can Democracy Survive the Internet?', *Journal of Democracy*, 28(2), pp. 63–76.

Schradie, J. (2019) *The Revolution that Wasn't: How Digital Activism Favors Conservatives*. Cambridge, MA: Harvard University Press.

Walker, J.L. (1991) *Mobilizing Interest Groups in America: Patrons, Professions, and Social Movements*. Ann Arbor, MI: University of Michigan Press.

12 Symbolic politics meets digital media: research on political meaning-making

Lone Sorensen

Introduction

In the hit Ukrainian TV political satire *Servant of the People*, a teacher becomes president when his expletive-laden rant about the deceitful state of Ukrainian politics goes viral. On 21 April 2019, Volodymyr Zelenskiy, who played the character of the accidental president, turned fiction into reality. Zelenskiy was elected president of Ukraine with an unprecedented 73 per cent of the vote. In the run-up to the election, Zelenskiy hardly engaged with mainstream media. Instead he posted informal video blogs online that he filmed on his smartphone. He openly admitted his lack of political views, just like his innocent teacher character. Indeed, he capitalized on his political ignorance and crowdsourced his political programme on social media. This played into his rather vague slogan of 'The president is a servant of the people', which in turn echoed the title of his TV show.

A distrustful citizenry knows that politicians' mainstream media appearances are rehearsed and staged. Zelenskiy's unorthodox avoidance of such put-on mediated performances and reliance on digital media was a symbolic act that played with truth, fiction and reality. It was a disruption of the media–politics nexus that constituted a claim to authenticity and a refusal to play the game of dirty politics. The idea, he said, was that people wanted 'a person with a human face' (Fisher, 2019). Ironically, an actor-in-character could satisfy this desire with more apparent honesty and sincerity than any politician by augmenting his authenticity through digital media. 'This isn't fake,' he insisted in one of the only media interviews he did (Fisher, 2019).

But what happened in that election is not unique to Ukraine. Fake politics is exactly what concerns many ordinary voters in liberal democracies today. Although the trend is not entirely uniform, most surveys show a decline of public trust in mainstream politicians and an increase in voter apathy around the globe (Duncan, 2018: 1–2; Norris, 2011). And, like Zelenskiy, political actors respond to the desire for authenticity and disruption of the fake politics-as-usual. Donald Trump's 2016 United States (US) presidential campaign contrasted the political newcomer with 'crooked Hillary' and the fake liberal establishment. He diverted press briefings to Twitter to avoid 'fake' establishment media. Italy's Five Star Movement began in 2009 as an online platform where the politically disaffected could air grievances against the political system. And Occupy Wall Street likewise positioned themselves on the periphery of institutional politics through their 'We are the 99 per cent' meme. Their use of digital technology channelled public antipathy to a fake establishment by allowing citizens to express individual feelings of injustice (Bennett and Segerberg, 2013). In all these cases, and many more, digital platforms furthered symbolical challenges to the falsity and oversight of self-serving establishment politics.

Performance theory and symbolic meaning-making

The preoccupation with fake politics grows out of a public concern with political meaning-making. People are increasingly questioning the relation of politicians' statements to reality, and to the speakers' own beliefs: are politicians basing their ideas on comprehensive and reliable evidence, and are they being true to themselves? Looking at politics through the lens of political performance can do more than point a finger at the spectacle and supposed deceit of modern politics. Beyond this everyday derogatory sense, political performance is a theoretical perspective that allows us to understand power through struggles over symbolic meaning-making.

Edelman (1985) coined the phrase 'symbolic politics' to foreground political symbols as the chief means through which the public knows and engages with political situations. Citizens' propensity to orient on the symbolic rather than the factual side appears to have increased even further in recent years. Alexander's (2006) theory of social performance goes one step further than Edelman and reverts the relation between power and meaning-making: it conceives of cultural-symbolic struggles as determinants of power rather than of meaning-making as in the service of power. The encoding and telling of stories enable control and create social structure. This strong theory of meaning-making adopts a more multidimensional approach than Edelman's exploration of an ostensibly one-way manipulative communication process: it sees performative habitus as potentially anybody's.

Performance theory also takes account of the interplay between the material and the non-material qualities of communication. The material qualities of intersubjective exchange can here be understood as the embodied ways in which people show that they do things, the objects they use to show this, and restrictions on access to these means of symbolic production. The non-material properties are culturally specific background representations, the narratives and cultural codes that performers draw on to encourage recognition, identification and emotional engagement in the audience. Performance theory approaches symbolic meaning-making as a collaborative enterprise that sees performances develop between actors and audiences in ways that allow the audience to suspend disbelief and invest in the actor's character. Or not, as in the case of former United Kingdom (UK) Prime Minister Theresa May's lack of performative habitus in her 'Abba moment' when she awkwardly danced onto the Conservative Party Conference stage in October 2018. A theory of political performance allows us to approach the process of meaning-making where it really matters, namely at the point where politics emerges. This is the point at which embodied acts, directed at an audience as if on stage, are imbued with new meaning, and a disruptive optic constructs political reality. This is where political agents engage in struggles over meaning and truth and seek to define our political existence in collaboration with the audience.

Zelenskiy's campaign is one of many examples of political meaning-making that is taking digital and hybrid forms where these forms in and of themselves symbolically convey meaning. But to study such political meaning-making in the hybrid media system (Chadwick, 2013), there are three key issues that scholarship needs to address. First, studies of digital politics largely go in one of two directions. Analysis of political content is often based on the assumption that the platform of communication is irrelevant to the meaning of its content. Conversely, big data investigations into social influence or structures of opinion concern themselves exclusively with instrumental uses of platform infrastructure (in essence, who clicks on what to achieve a specific goal). Both miss the symbolic aspect of meaning-making whereby digital artefacts are used to symbolically create meaning that is then anchored in content. An extension of political performance theory can bridge these perspectives and address the gap of symbolic meaning-making.

Second, most political communication research confines itself to silos of media types or platforms (Bode and Vraga, 2018) that do not reflect the hybridity of political media usage. I suggest that the notion of performative assemblages comprised of acts mediated through, and aimed at, a variety of media can usefully be adopted for this purpose. Third, digital communication creates strong emotionally based relationships between representatives and supporters. We need to turn our attention to different types of actors' means of achieving this, and what such engagement means for political efficacy and participation. I address each of these issues in turn, and suggest avenues for a more expansive research agenda that develops and extends the theory of political performance and associated methods to accommodate the virtual stage. I then outline some methodological principles that can guide how we may go about achieving this.

Digital symbolic action

Edelman's argument that people engage more with politics through the symbolic than through concrete and manifest struggles over resources between opposing interest groups, also applies to digital politics. Increasingly, political actors and organizations, such as Spanish Podemos and Jeremy Corbyn's Labour Party and its associated Momentum movement in the UK, borrow digital communication strategies from social movements. They mobilize on social media by performing symbolic acts aimed at generating emotion and constructing community (see e.g. Chadwick and Stromer-Galley, 2016 on parties-as-movements; Gerbaudo, 2014 on populist actors; Treré, 2018 on political activists). They ascribe the technical infrastructure of digital media and the actions it affords – multi-way communications, mass publishing, and so on – with meaning by tapping into the ways in which these technologies are imagined as, for example, democratic, non-hierarchical and anti-elitist. An online poll of political supporters, for example, may not be intended to gather views but to make people feel listened to. The utilization of media imaginaries – the ideologies, values, meanings and imagined capabilities of technology – in turn shapes political practice (Treré et al., 2017).

Yet most digital research fails to attend to symbolic politics (and most research on symbolic politics neglects the digital). An extension of performance theory could attend to this neglect, as digital media offer parallels and extensions to both the material and the cultural dimensions of performance. Social power grants or restricts performers' access to media, creates the regulations that govern digital activities, and designs media artefacts that afford certain acts, such as clicks and posts, to mass audiences (Isin and Ruppert, 2015). Such 'digital acts' (ibid.) in turn tap into culturally specific background representations through their content but also through media-specific imaginaries to act out scripts and create meaning that audiences, in the form of 'prosumers' (Toffler, 1984), take a more active role in reshaping and distributing online.

Zelenskiy's reliance on social rather than gatekept media in his campaign can, from this perspective, be seen as symbolic as much as instrumental. At one point, Ukrainian establishment politicians allegedly spread fake news stories about Zelenskiy through mainstream media to undermine his campaign. In response, Zelenskiy announced an online competition to invent the most creative fake news about himself, and publicized the satirical offerings online: he was really a metamorphosed reptile; he was descended from the Rothschilds; he was dating Angela Merkel. Not only was the 'real' fake news exposed and ridiculed, but the online competition also enabled the public to actively engage in the revelation of the establishment's deceptive use of legacy media. The harnessing of imaginaries of anti-elitism and people power associated with the direct link between candidate and people that digital media afford became powerful means of symbolic production in Zelenskiy's performance. The symbolic creation of fake news became a claim to truth.

Extending performance theory to the digital can address how claims to representation and truth are made through digital media; the extent to which symbolic claims translate into real political listening, participation and efficacy; whether the mechanisms of mobilization and engagement on social media enable the spread of disinformation, hate speech and fear as well as connective and collective action (Bennett and Segerberg, 2013); and under what conditions this is the case. These are essential questions for the study of politics in the digital age and can only be addressed by an approach that considers the instrumental and material uses of media in combination with their symbolic function.

Performative assemblages in the hybrid media system

The silos of media types, whether digital or analogue, that currently confine much political communication research prevent a proper understanding of the role that digital media play as part of larger and more complex communication repertoires. Political performances are nearly always assemblages that are acted out through, and directed at, multiple media platforms (Sorensen, forthcoming). They integrate live action and social media and are targeted at both supporters and traditional gatekeepers in increasingly elaborate ways.

The digital elements of Zelenskiy's campaign were designed as a reality TV show, with online broadcasts of the work of his campaign office and impromptu selfie videos as he went about his day. Zelenskiy's digital performances were a symbolic rejection of the polished and contrived presentation that characterizes mainstream media interviews. Instead they signalled unfiltered transparency and unstaged authenticity. Indeed, he declined practically all media interviews, talk shows and TV appearances, prompting 20 news outlets to ask him to stop avoiding journalists in an open letter. He did not, however, reject mainstream media: only its gatekeepers. For instance, he often included snippets from his TV show in his video blogs. The popular 1+1 TV channel premiered the latest series of *Servant of the People* four days before the first-round vote and dedicated the full evening before the election to reruns of Zelenskiy's shows. Zelenskiy also channelled his on-screen persona in live rallies. Rather than giving political speeches, he would appear in character on a stage set up as a circus and deliver a stand-up comedy show.

These performances were hybrid in another sense than using multiple and interdependent media forms. They were playing with the boundary between fiction and reality. Zelenskiy-the-candidate merged with his TV character and performed a message of transparency and authenticity through a fictional TV creation. Yet the voting public felt this fiction to be more truthful about Ukrainian politics than any factual media report or political interview given by establishment contenders. Only by approaching the campaign as an assemblage of complementary media performances where each channel contributed its own affordances and imaginaries can we dissect Zelenskiy's approach to political meaning-making that inspired so many discontented citizens.

Media scholars are well aware of the integrated nature of the hybrid media system. Yet many dominant theoretical perspectives in political communication – for instance, on media logic, political participation or audiences – are confined to particular media types. Behind this lies a hesitancy in overcoming the long-standing distinction in media and communication studies between mass and interpersonal communication, which dissolved with the advent of social media and has become increasingly complex as the media environment has hybridized.

Another reason for the persistence of media silos is a reliance on platform-specific methods and tools in empirical research. Technical restrictions (or too-convenient solutions) end up driving research questions away from the reality of hybridity. They stop us seeing how meaning travels and changes across platforms and between online and offline environments, or how political actors construct, and citizens receive and interact with, messages and meanings in a multitude of channels. A performance perspective can encompass institutionalist concerns with legacy media as well as the network logic of digital media. It can help us to consider how the material affordances, norms and cultural imaginaries of any media are used symbolically and complement each other in the construction of meaning through hybrid performative assemblages. Before I address the glaring methodological issues that hybridity raises, I want to suggest a third aspect of digital political meaning-making that ought to be on our agenda.

Emotional engagement

When we delve into political conversations and social relationships, we usually find that they are emotionally driven. Emotion is dismissed by most rational-world accounts of democracy as undermining reasoned decision-making. With the advent of digital politics, the polarizing tendencies of social networks and unrestrained exhibition of vitriol online have become major concerns. Yet any construction of identity or of a people requires affective investment. As such, emotions are legitimate evaluative responses to performances of political meaning (Coleman, 2013). Asking how we can exterminate the viral spread of emotions may therefore not only address the worrying trends in digital politics but also undermine democratic practice and engagement altogether. Instead, we need to consider how some political actors engender feelings in audiences for purposes other than those openly professed by their performances; what the implications are of emotionally distancing people from their representatives by exposing the latter's theatrical artifice; and how we can reinvigorate representative politics with public feelings of efficacy.

Social performance literature tells us that identification between dramatic characters and audiences creates an emotional bond (Alexander, 2006). The foundation of this relationship rests upon the audience's suspension of disbelief: they willingly forget the distinction between character and actor. A political performance perspective therefore invites exploration of the conditions under which citizens are willing to suspend disbelief – in Zelenskiy's case, despite his professed fictional self – and when the bond is broken. It is a lens through which to examine the growing harnessing of emotion and development of visceral intersubjective understanding that is established on the stage of digital platforms.

Zelenskiy's crowdsourced policy platform tapped into Web 2.0 imaginaries of non-hierarchical power structures, community-building and direct participatory democracy to achieve this. While he received many hundred suggestions through his digital platform Lift, the membership – a requirement to submit a suggestion – currently stands at over 620 000 (July 2019). The intimate community feeling engendered by the protective virtual walls of online group membership should not be underestimated, neither should the force of symbolic listening, which reaches far beyond members. We saw a similar surge of communal triumph in the 2016 US Democratic primary at Bernie Sanders' ability to fundraise through $1 online donations in defiance of the special-interest influence of large-scale campaign donors. Like Sanders, Zelenskiy generated an affective bond with and within the electorate. By combining virtual community-building with a projection of himself as authentic, ordinary and relatable, in contrast to a political establishment portrayed as deceptive, he turned Ukrainian politics upside down.

Existing research on emotion and digital politics focuses on othering, polarization and hate speech. Beyond these topics, the performance of emotion in digital politics is still a relatively untouched and ripe area for investigation (but see Karatzogianni and Kuntsman, 2012), and one that is increasingly pertinent. But we are faced with a methodological obstacle. Studies of polarization mostly rely on social network analysis to demonstrate the social structures of emotional discourse, but we know very little about how such structures come about through political meaning-making. Sentiment analysis can never hope to capture the 'how' of emotional performance (and so far only very imperfectly the 'what'). Digital emotional engagement, like the hybrid nature of political performances, requires us to put methodological development at the forefront of our research agenda.

Methodological principles

How do we empirically study the complex processes of meaning-making in the hybrid media system? The continuous evolution of the media ecology, and the complex constructions of political reality, call for adaptation and sustained flexibility from our established and proven methods. I do not here propose a ready-made solution, but rather a set of methodological principles that can guide further collective endeavours.

Paying attention to agency in social change

Performative assemblages, if successful, can become critical events that change the course of history or serve to maintain the status quo. A performance perspective predicates the creation of change at the macro level of social structure upon agency at the micro level of social interaction. In other words, it conceives of social change as a cultural and performative process (e.g., Roudakova, 2017). Yet performance theory also recognizes that social power grants or restricts access to the means of symbolic production and consequently affects the impact of a performance (see Figure 12.1). To address this duality, critical events analysis (Kraus et al., 1975: 195–7) suggests that we integrate micro-level studies of individual performers and performances with macro-level data at the societal level of analysis.

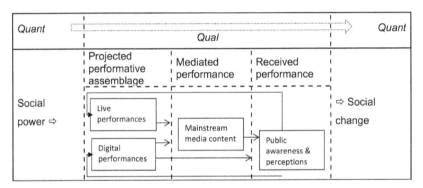

Figure 12.1   Performance-centred data-gathering in the hybrid media environment

Mixed methods

Macro-level data can inform exploratory and interpretive analysis in three key ways. It can create causal models; it can provide context for interpretations of individual performances by testing the relevance of structural factors for performances at the aggregative level; and it can form an initial pool of data on a delimited performance (see below) for further qualitative sampling. Social network analysis may, for example, measure the level of polarization online under different conditions, but it cannot explain how it has come about. In a mixed-methods approach, digital infrastructure and measurements of big data meet 'thick data' (Wang, 2013), which reveal the social context and stories behind data points. Zoom-in, random or top sampling for qualitative analysis are means of getting at these deeper structures of meaning (Gerbaudo, 2016).

We can then start to understand how social power and media systems, infrastructures, norms and practices constrain, enable or enhance particular aspects of individual or group performances, and how such performances have social consequences.

Delineation of the performances under study

Performances, like events, are difficult to delineate, and a choice of a particular beginning and end can change their meaning (Wagner-Pacifici, 2010). A perspective that focuses on agency as the driver of change and performances as intentionally designed can follow the key actors (Marcus, 1995) in sampling operations, letting their own engagement with an event guide the delineation of the performance.

A media ecology approach to sampling

Engagement with political meaning-making in the hybrid media system calls for a holistic approach that acknowledges the multi-site, context-dependent and responsive construction of meaning. Simply turning our gaze to Twitter and adopting the latest handy tool for data collection and visualization does not do the trick, for Twitter is not where political reality is exclusively constructed or experienced. Instead, I suggest that an event-based approach can guide data collection. Kraus et al. (1975: 210–212) outline three basic data-gathering operations, which I have here adapted to a multi-site hybrid media performance (Figure 12.1). First, reconstruct what happened, what the goals of key actors were and how the media operated. This involves a study of the performances projected by political actors, live and online, and interviewing key political and media actors. Second, compare the performance mediated through mainstream media to the projected performance. Third, gauge the public response to, understanding of and involvement in the performance in relation to different channels. Public interviews, reactions by platform-specific audiences such as Twitter, and opinion polls can be useful data sources for this purpose. Mapping these multi-site performances within and across media can then identify critical and contentious instances of meaning-making.

A process-oriented mode of analysis

The concern with the role of cultural process in social change suggests engagement with the story in its telling, not its narrative or themes. Grounded theory's (Charmaz 2006) approach to process coding and analysis involves constant comparison within and across cases and contexts, and sets a performance within the full social process in which performers link to audience reaction and social change. It lends itself to consideration of context at a number of levels, including relevant social conditions and patterns identified at the macro level, the materiality and imaginaries of individual media technologies, and the larger conversation within and outside of the given media platform of specific data points.

Conclusion

A decline in public efficacy, and preoccupation with politicians' fake performances, appear to be spiralling developments. Mainstream news media have in recent decades become increasingly preoccupied with political strategizing and what they pejoratively term 'political theatre', rather than the content of policies (Blumler and Gurevitch, 1995). Public awareness of politicians putting on a performance then shows itself in declining efficacy and growing apathy. In response, political actors brand themselves as the new authentic thing and seek to gain political capital from exposing the false façade of establishment politics. By combining authentic self-representation with disruption and popular culture, some manage to simultaneously respond to the media's demand for spectacle and engagement with citizens as audiences. Personalization, populism, peripheral and symbolic politics all speak to this trend. At times, such performances of authenticity are themselves exposed as acts; and so, it seems, we continue the downwards spiral.

Zelenskiy's campaign performance added another layer to the fake politics script. His performance of authenticity was explicitly, openly and honestly a performance. He did not pretend not to pretend; instead he appeared on stage in character. His performance thereby became a symbolic commentary on the theatre of establishment politics, and he himself became a symbol: 'Zelenskiy the Ze'. He achieved this through a hybrid performance. Refusal to participate in the media circus of mainstream campaigning enabled identification with his TV persona. This was reinforced by the direct link to his supporters, who were able to actively participate in his campaign online. But the real Zelenskiy is not identical to his innocent TV character. He is a millionaire and TV network owner with suspected links to one of Ukraine's most prominent oligarchs who, incidentally, owns the TV channel that aired his shows the night before the election. Zelenskiy's was a performance choreographed for the hybrid media system and its changing relationship with truth and fiction, and so far it has proven impressively effective and efficacious.

To gain a proper understanding of these new and changing trends in digital and hybrid political meaning-making, I have suggested an agenda of three connected research areas that are woven together by a concern with political meaning-making. The first is one that marries the imaginary and material dimensions of meaning-making to pay attention to symbolic action. The second considers how such communicative practices play out across the hybrid media system through performative assemblages. The third examines how they engage emotionally with audiences. I have advocated a set of principles as an opening gambit for further methodological work to address these areas. These principles are informed by a theoretical perspective that places agency at the centre of social change, and sees the interaction between material and imaginary qualities of communication as the crux of a symbolic meaning-making process. The study of digital political performance observes physically embodied acts and the affordances, norms and uses of digital infrastructure. But it also looks beyond the material. The approach encourages us to address the ways in which political agents purposely associate material things, practices and norms with recognizable imaginaries and background representations to create meaning. Such symbolic meaning-making is always new and contentious in its construction of reality: that is the nature of meaning-making. But in the current moment, its most successful manifestations are dismissing the charade of politics in favour of new forms of truth as popular and political communication merge to reinvent the representative relationship. And that makes understanding it critically important.

References

Alexander, J.C. (2006) 'Cultural Pragmatics: Social Performance between Ritual and Strategy'. In J.C. Alexander, B. Giesen and J.L. Mast (eds), *Social Performance: Symbolic Action, Cultural Pragmatics, and Ritual.* Cambridge, UK and New York, USA: Cambridge University Press, pp. 29–91.

Bennett, W.L. and Segerberg, A. (2013) *The Logic of Connective Action: Digital Media and the Personalization of Contentious Politics.* New York: Cambridge University Press.

Blumler, J.G. and Gurevitch, M. (1995) *The Crisis of Public Communication,* Communication and Society. London: Routledge.

Bode, L. and Vraga, E.K. (2018) 'Studying Politics Across Media', *Political Communication,* 35, pp. 1–7. https://doi.org/10.1080/10584609.2017.1334730.

Chadwick, A. (2013) *The Hybrid Media System: Politics and Power.* New York: Oxford University Press.

Chadwick, A. and Stromer-Galley, J. (2016) 'Digital Media, Power, and Democracy in Parties and Election Campaigns: Party Decline or Party Renewal?', *International Journal of Press/Politics,* 21, pp. 283–93. https://doi.org/10.1177/1940161216646731.

Charmaz, K. (2006) *Constructing Grounded Theory: A Practical Guide through Qualitative Analysis*. London, UK and Thousand Oaks, CA, USA: SAGE Publications.

Coleman, S. (2013) *How Voters Feel*. New York: Cambridge University Press.

Duncan, G. (2018) *The Problem of Political Trust: A Conceptual Reformulation*. London: Routledge.

Edelman, M.J. (1985) *The Symbolic Uses of Politics*. Urbana, IL: University of Illinois Press.

Fisher, J. (2019) 'What Does a Comic President Mean for Ukraine?', *BBC News*, 22 April. Available at https://www.bbc.co.uk/news/world-europe-47769118 (accessed 11 September 2019).

Gerbaudo, P. (2014) 'Populism 2.0: Social Media Activism, the Generic Internet User and Interactive Direct Democracy'. In D. Trottier and C. Fuchs (eds), *Social Media, Politics and the State: Protests, Revolutions, Riots, Crime and Policing in the Age of Facebook, Twitter and YouTube*. New York: Routledge, pp. 67–87.

Gerbaudo, P. (2016) 'From Data Analytics to Data Hermeneutics: Online Political Discussions, Digital Methods and the Continuing Relevance of Interpretive Approaches', *Digital Culture and Society*, 2, pp. 95–112. https://doi.org/10.14361/dcs-2016-0207.

Isin, E. and Ruppert, E. (2015) *Being Digital Citizens*. London, UK and New York, USA: Rowman & Littlefield International.

Karatzogianni, A. and Kuntsman, A. (2012) *Digital Cultures and the Politics of Emotion: Feelings, Affect and Technological Change*. Basingstoke: Palgrave Macmillan.

Kraus, S., Davis, D., Lang, G.E. and Lang, K. (1975) 'Critical Events Analysis'. In H. Chaffee (ed.), *Political Communication: Issues and Strategies for Research*. Beverly Hills, CA, USA and London, UK: SAGE Publications.

Marcus, G.E. (1995) 'Ethnography in/of the World System: The Emergence of Multi-Sited Ethnography', *Annual Review of Anthropology*, 24, pp. 95–117. https://doi.org/10.1146/annurev.an.24.100195.000523.

Norris, P. (2011) *Democratic Deficit: Critical Citizens Revisited*. Cambridge: Cambridge University Press.

Roudakova, N. (2017) *Losing Pravda: Ethics and the Press in Post-Truth Russia*. Cambridge, UK and New York, USA: Cambridge University Press.

Sorensen, L. (forthcoming) *Mediated Populism: Performance, Ideology, Hybridity*. Palgrave.

Toffler, A. (1984) *Future Shock*. New York: Bantam Books.

Treré, E. (2018) *Hybrid Media Activism: Ecologies, Imaginaries, Algorithms*. New York, USA and London, UK: Routledge.

Treré, E., Jeppesen, S. and Mattoni, A. (2017) 'Comparing Digital Protest Media Imaginaries: Anti-Austerity Movements in Greece, Italy and Spain', *TripleC*, 1(15), pp. 404–22. https://doi.org/10.31269/triplec.v15i2.772.

Wagner-Pacifici, R. (2010) 'Theorizing the Restlessness of Events', *American Journal of Sociology*, 115, pp. 1351–86.

Wang, T. (2013) 'Big Data Needs Thick Data', *Ethnography Matters*. Available at http://ethnographymatters.net/blog/2013/05/13/big-data-needs-thick-data/ (accessed 9 September 2019).

13 Sending a message: the primacy of action as communication in cyber-security

Ben O'Loughlin and Alexi Drew

Introduction

Action is the primary form of communication in cyber-security. In a field in which the capacities and intentions of actors are uncertain to others, nothing sends a message like an attack. Unlike electoral campaigns, digital diplomacy or even the public legitimation of warfare, it is not a realm of proactive story-telling. This means that researchers must devise frameworks to interpret the meaning of action. Efforts are made in national and international forums to establish laws and norms for cyber-security. These are secondary, however. They lag behind, and are reactive to, action.

Nevertheless, communication around laws, norms and the meaning of actions remains important and we recommend this as a focus of research. In particular, the last decades have seen a dialectical movement between politicians seeking to fix the Internet as an object of security policy, which might be called 'securitization', and efforts by other political actors to free the Internet from that framing, which has been called 'de-securitization'.

This dialectic between securitization and de-securitization could have important consequences. If the Internet and digital technologies more generally become understood as a terrain of cyber-warfare and cyber-terrorism, then securitization serves to authorize a military response. It casts the Internet as a matter of national security in ways that justify different policies than if it was governed as a civilian or civic technology.

In this chapter we unpack how action and communication about action operate around cyber-security. We propose a set of considerations for researchers to develop frameworks to analyse action and its meaning. We highlight the urgent need for scholars to map how cyber-security operates comparatively across countries, because without an understanding of national and international architectures of cyber-security we cannot make valid claims about what a realistic digital politics of cyber-security might be.

Finally, we set out a series of agenda points for future research. Cyber-security as a field within digital politics is extremely challenging. It requires numerous skill sets and imaginative workarounds to open up exactly who is doing what to whom. It necessitates extreme sensitivity to the manner in which any concepts agreed upon in academia and ultimately international law will then be exploited by political actors. These challenges make cyber-security one of the most exciting and significant areas of research in digital politics.

Action and communication: normalizing conflict

In June 2019 the United States (US) engaged in an open, targeted attack on Russian critical infrastructure, specifically the power grid, in response to the continued Russian interference in US domestic processes such as the 2018 mid-term elections. Later that same month the US engaged in a similar campaign against a more opaque set of targets in Iran as part of a package of responses to the Iranian downing of a US Global Hawk drone. These actions are relatively new and obvious contributions to a process which accumulates normative legitimacy through repeated operationalizing of the ideals which it represents. By engaging in cyber-attacks against the critical national infrastructure of two separate states the US is contributing to a normative position legitimizing this digitized use of force between state adversaries. This normative position has been previously contributed to by one of the targets of the US's most recent attacks. Russia's cyber-attacks targeting the power grid of Ukraine in 2015 and 2016 further advance, through action, the normative standard that cyber-attacks between states are an acceptable practice.

Evidence of action driven, operationalized construction of norms within the broad web of the cyber domain can also be observed in more domestically focussed normative concepts. Cyber-surveillance, the umbrella term for the investigation of individuals using evidence gathered from their digital footprint of use of online services, is becoming a powerful tool in the eyes of security services worldwide, and a pernicious threat to individual freedoms for civil rights groups. Ethical and legal arguments surrounding the United Kingdom's retroactive extending of surveillance powers in the digital space through the Investigatory Powers Act, coming on the heels of Edward Snowden's revelations of state surveillance, have since proliferated in global politics. Chinese surveillance of its citizenry, particularly of ethnic minorities such as Uighur Muslims, epitomizes the dangerous nexus point of technological advance and the drive to provide security at any cost. In China, much as within the United Kingdom (UK), justifications for this extension of powers employ both frameworks and language that stress the need to provide security from a distinct and existential threat: terrorism.

The UK has, alongside the Investigatory Powers Act (Her Majesty's Government, 2016), updated their definitions of terrorism with legal frameworks to include references to cyber-attacks. Politicians have latched on to terms such as the 'dark web', and civil servants with a security background have decried technology platforms as the 'command-and-control platforms' of terrorists. There remains little evidence of acts of cyber-terrorism that might encapsulate the existential risk to the level at which they are portrayed in this discourse. Cyber-attacks by groups such as the Syrian Electronic Army, or Hamas, have had limited impact. And yet both victims and perpetrators of such attacks have sought to amplify their importance through words and deeds. One such response has seen Israel targeting the base of operations for CyberHamasHQ.exe, with a kinetic military strike destroying a building in Gaza in June 2019.

Actors launching cyber-terrorism attacks can be called 'norm entrepreneurs' because they set the agenda for what types of action are to be regulated and what news media will characterize as a cyber-terrorism attack. Like states launching cyber-attacks, terrorist groups use violence to achieve goals, and must balance opportunity and risk. As with the Irish Republican Army (IRA) statement to the UK, 'We only have to be lucky once. You have to be lucky every day' (*BBC News*, 2005), so cyber-terrorism can create fear about potential attacks. It is an empirical question to ask: is uncertainty about the form and consequences of such attacks produced as much in the imaginaries of societies fearing attacks as by terrorist groups themselves? News media and popular culture can amplify anxieties about terrorist threats.

Political leaders may amplify threats in order to justify more hardline security policies and extensions of existing powers. This can produce a reinforcing process as the interests of both terrorist groups and governments converge. As terrorist groups amplify terror through operational and rhetorical acts, states may inflate rhetoric and may even launch and publicly acknowledge counter-terrorism cyber-attacks against those groups. Researchers tracing these inadvertent reinforcement effects over time can build explanations of how expectations and norms are produced through action and counter-action and the attendant rhetoric. In the 2010s it has been evident that terrorist groups launch attacks, and states react to and amplify them, thereby embedding as the norm the action the terrorist group created.

If states feel the need to launch cyber-attacks on rivals in a manner that undermines or subverts existing norms or international law, it indicates that those states feel constrained by the international system and its standards of behaviour. They act in their interests, from their perspective, by contravening norms and law. Digital politics scholars may be able to analyse the narratives of state leaders, military officials or intelligence officers to detect when they feel current limitations on their behaviour begin to chafe. The US and Israel demonstrated this with Stuxnet, having threatened to sabotage Iranian facilities in the years prior to the attack. The UK has bridled against norms when threatening cyber-retaliation against Russian-attributed poisonings.

The aesthetics of action

It becomes necessary to have an approach for researching action. There is an ambiguity to so many manoeuvres undertaken in the terrain(s) of cyber-security. An action such as Stuxnet is a direct attack by states on another state, with material consequences. Iran puts great effort into maintaining the anonymity of its nuclear scientists and maintaining the secrecy of its militarized facilities, while allowing enough transparency of its nuclear facilities to display to the global community that it is a peaceful and cooperative nation-state. To penetrate, corrode and shut down Iran's nuclear facilities is therefore a gesture, designed to be felt. When Israeli officials were asked whether they were the perpetrators of Stuxnet, they grinned broadly (Broad et al., 2011).

The gestures or performances are one potentially rich avenue for research in the digital politics of cyber-security. This could build on a tradition of studies of the ways in which the conduct of international security is highly gendered, not just in the labels used for weapons or to describe actions towards allies and enemies ('we took them', 'we nailed them'), but in the ways such discourse produces particular understandings of agency, nationhood, sovereignty, and what counts as security and to whom (Cohn, 1987). This opens the field to visual and critical analysis as well as conventional policy process-tracing and analysis of language. In the international relations (IR) field, Brent Steele

(2010) has argued that power invokes the need for its own display. His analysis of US foreign policy shows how a great power nation-state must demonstrate its power. That power must be witnessed, and there must be a sense that others feel the power.

In cyber-security such actions beget reactions: political leaders can relish disrupting norms, and then appear horrified and appalled by opponents' action. Cyber-attacks, whether by state or non-state actors, allow for the performance of technical proficiency and symbolic destruction. Analysis of such performances opens up the field to feminist analysis of the gendered displays of mastery or strength and their subversion. This allows for attention to the emotional dimensions of cyber-security communication, such as displays of pride, wounded honour and even trollish abandon. Such research would be significant because even conventional, variable-based analysis of international security demonstrates that a state's gender attitudes predict its propensity to war more than any other factor (Hudson et al., 2012).

The digital politics of securitization: who buys it?

Norm construction through action, and the subversion of existing norms, allow states to make gains in terms of prestige and presence while offsetting the risk of being 'called out' in a manner that would lead to severe repercussions from the international community. Discourse and norm-oriented efforts towards reaching global consensus on cyber-security behaviour have been undermined by a failure to anticipate what political actors could do through cyber-attacks that are ostensibly non-violent acts. Those efforts have also been undermined by the actions of states which offered rhetorical support to those norms. All sides say that they support international norms and law, yet all sides launch cyber-attacks. Attacks such as Stuxnet and Flame and others effectively cast doubt on the credibility of that rhetorical support. This suggests that operationalized norms, or what Dunn Cavelty calls 'behavioural norms', are proactive and agenda-setting, whereas their discursive cousins are reactive and constantly seeking to redraw and rein in the boundaries of appropriate behaviour (Dunn Cavelty, 2013: 115).

Norm entrepreneurs seeking to drive a process of de-securitization find themselves at an automatic disadvantage. This is due to the combination of powerful emotive language implicit in the discourse-driven normative polishing which is engaged upon by securitizing norm-entrepreneurs, and the operationalization of these securitized norms through action embarked upon by these same securitizing entrepreneurs. An effective de-securitization would require not only an effective countering of the language used to frame the threat as imminent and existential, undermining the risk posed to often emotively selected groups (such as children), and simultaneously an operationalizing of the preferred, desecuritized norm through definite positive and punitive actions. The bar is set much higher for norm entrepreneurs who seek to restrain a trend of securitization in cyber-space.

The cyber-surveillance example provides both an instance where desecuritization has enjoyed some amount of success, and a case which demonstrates how the study of this phenomenon might be advanced in future work. The UK has successfully maintained the securitized norms of surveillance, based upon overexaggerated scenarios of operationalized norm construction through acts of cyber-terrorism and effective use of language that described existential and immediate risk. Europe, on the other hand, has managed significant progress in supplanting the securitized normative position. It has set itself up as a leading actor in the formulation of legal frameworks based on a framework of cyber-norms that sets individual rights above the needs of security. The General Data Protection Regulation (GDPR) framework and concepts of individual data ownership are but two of a growing list of examples. Each of these cases provides a direction for further research into the process of de-securitization. What norm entrepreneurs can be identified? What kind of language was used to supplant national security concerns with the concerns of individuals? These are all questions for future research to better understand, and hopefully export, the successful de-securitization of cyber-surveillance in Europe.

One particular facet of this research would be to examine how audiences of securitization – the population groups connected to the state or institution embarking on this route – react to both securitization and de-securitization. What are the sources of public opinion or feelings with regard to their risk from cyber-terrorism or attack? Is there significant variation between states, regions, economic or social backgrounds? This is an area of knowledge that has only begun to be examined by the academic and political community. It is one which could be central in undoing the damage caused by securitization processes which have relied heavily on the manufacture of a state of fear in its key audiences.

Inroads into these questions have been made. The examination into the Canadian news media's framing of cyber-terrorism on publics (Keith, 2005) sought to avoid conclusions as to the reality of the threat posed by cyber-terrorism, but instead to investigate how publics' and policy-makers' understandings are being shaped. They found that at the fundamental level the confusion as to what constituted cyber-terrorism in the academic space was amplified significantly in the journalistic one, and this led to an unclear and potentially inaccurate framing of the actual risk posed to the public audience.

Efforts to examine the knowledge of cyber-security issues by public audiences have included studies as to relative risk when balanced against other potential threats, such as crime.[1] Practical understanding of cyber-security issues still appears to be relatively low in the public sphere, suggesting that there is a gap with regard to the translation of information from expertise to general public. In a study of US citizens by Pew in 2017, only 1 per cent of respondents got a perfect score when asked 13 questions on cyber-security issues (Olmstead and Smith, 2017). The sources of this gap offer a ripe area for further research. This will be fundamental in efforts to ensure that those at risk from cyber-security issues are properly informed, even as these issues become more and more prevalent.

Mapping the global architecture of cyber-security (or its absence)

In so many areas of global politics, scholars lack a holistic understanding of the most urgent issues; and it is not always clear that policy-makers see or know more. Since the 2008 global financial crisis only a few scholars identified this as a moment to simply map how financial institutions and practices operate around the world. Regulators did not know; elected representatives did not know; the programmers of trading algorithms did not always know. This was a task that Adam Tooze (2018) set himself. He argued that only when we can see how global finance works can we realize the problems, and form positions, coalitions and alliances to face them. Until we know how the system works, then it is difficult to know whether a new idea or narrative is realistic or novel. The same applies to climate change, health and food security and, we would argue, cyber-security. We cannot have an intelligible digital politics of cyber-security until we understand what cyber-security is, what structures and systems exist in and across different countries, and who is steering the development of these technologies.

Researchers must account for the specific national trajectories of approaches to cyber-security by comparing the policy systems and technological infrastructures of different nation-states. This will be challenging. The availability of information varies. Researchers must pay attention to tendering processes and possibly use freedom-of-information tactics to generate information about the role of private and public actors in cyber defence and offence. China must be researched, probably from the outside. A division of labour is required in this research sub-field because no single scholar or team of scholars will have the legal, technological and cultural knowledge to make sense of cyber-security globally.

It is only with that systematic knowledge that we can begin to ask more politically interesting questions, such as: why some states might be more resilient to attack than others; why some leaders in some political cultures may be more or less aggressive or cooperative in their cyber-security behaviour; and when and how certain images or narratives 'cut through' and make a difference to global understandings of cyber-security. Hiroshima, Mai Lai, Abu Ghraib – the creation of a form of conflict entails the creation of a form of conflict image.[2] Cyber-security's image is yet to arrive.

Future research

Here we set out parameters for future research on cyber-security and digital politics.

Clarity about forms of warfare

War, conflict, warfare, terrorism – each has its own academic literature. Each can be used in combination. A state might launch technical attacks against another while also using disinformation about the situation. A terrorist group could use hacking and malicious code to access a state's information and then exploit it publicly to justify its wider terror campaign. Scholars must be alert and make choices about the degree of holism required to answer their research question. The primary label used in public research will determine which agencies might respond, practically. There is a lot at stake in retaining conceptual clarity.

Comparative systems

As suggested above, researchers must explain the specific national trajectories of approaches to cyber-security. They can do this by comparing the policy systems and technological infrastructures of different nation-states, as well as decisions made by political leaders when attacks occur. At the time of writing, Western governments fear China would use Huawei and 5G contracts for nefarious purposes. If scholars cannot study China in detail, they can study how others interact with China and draw inferences about Chinese policy. The difficulty of acquiring comparative information about cyber-security policy and actions leads to the danger of the intentionality fallacy. Since scholars can observe outcomes, there is a danger of attributing intent to those who may benefit from such outcomes. However, scholars would need to find behavioural patterns that might better evidence intent with consistency.

Governance

Scholars must identify who sets laws and policies on cyber-security and who has the power to police those laws and policies. The creation of the United Nations Group of Governmental Experts on Advancing Responsible State Behaviour in Cyberspace in the Context of International Security (UN GGE) showed a recognition of the need to establish international norms. That it split into two bodies shows how difficult this will be. Nevertheless, this is an area where the methodological toolkit of politics and international relations can be useful. Scholars can conduct discourse and content analyses of these debates and treaties while producing policy analysis that captures the sequence of decisions and actions, tracing the interplay of language and behaviour over time (Krippendorff, 2017). For example, as we write in early 2019 it is clear that Russia wants sovereignty in cyber-space to stop others attacking it, but has no intention of stopping attacking others. Researchers must explore the temporal lags between what a state says and does, and produce contextualized inferences about states' choices of language use when discussing laws and norms.

AI

There are already numerous activist and policy working groups investigating the ethical and technical aspects of the role of artificial intelligence (AI) in cyber-security, such as around face recognition. There is a need to research how defence ministries are treating it. Since they are charged with assessing and meeting cyber threats, we could assume their assessments to be realistic. However, what they would publicly say about threat exposure is a sensitive political matter. There is also a need to research how AI is being developed in the private sector. For instance, in early 2019 Microsoft staff publicly refused to support a research project to develop technologies for the US military that might increase 'lethality' (Romm and Harwell, 2019). Such incidents generate policy documents, interviews and evidence of interactions between those actors and others in the defence sector.

Attacks or campaigns?

Researchers initially analysed individual cyber-attacks, but Valeriano and Maness (2015) have shown the utility of tracking operations or campaigns in dyads of states, for instance Russia–US campaigns of mutual cyber-attack and -defence. In these dyads we find continual testing of boundaries and breaches, and problems of signalling, attribution and proportionate use of force. There is no reason to expect these entwined relations between states to fade, and we need more dyadic analysis, as well as exploration of triads and other durable, antagonistic relations. There is also a need for conceptual clarity on what constitutes an 'attack' or a 'campaign'. This is difficult, because any aspect of society can be attacked. A state can promise not to attack another's electricity grid, but seek to undermine its social fabric through digital disinformation. It is also difficult because definitions have real-world consequences. If academia and international law reach consensus on what threshold constitutes an attack, then states will simply operate just beneath that threshold. A further, open question remains: can cyber campaigns ever end? As with the War on Terror, the boundaries of war, warfare and terror can blur, and 'war' can unfold without any clear start or end point. Cyber-security may be a realm of what electoral scholars are long familiar with: the permanent campaign (Blumenthal, 1982).

Projections and simulations

A range of actors construct expectations about how cyber-security will develop and what best- and worst-case scenarios can be imagined. Film directors create public imaginaries of how different kinds of breakdowns might unfold. Mainstream news media present risks through the use of frames and narratives, and their choice of expert sources as authoritative voices. Think tanks and universities publicize their simulations and wargames (Fitzpatrick, 2019; Shlapak and Johnson, 2016). Even the culture industry has noticed cyber-security: Escape Rooms in Prague, Sofia, Amsterdam and other cities offer a 'hacker room' for tourists to puzzle their way out of. Cyber-war unites two things that generate public anxiety – technology and conflict – and may also create fears of the total destruction and loss of technology.

Technological literacy

At present, it is striking how so many scholars advance different conceptions of 'cyber' because of different technical understandings. Researchers in the field of digital politics must complement their knowledge with engagement with information security, computing science and law. They must talk to those in the public and private sectors who are building these technologies. It may then be possible to achieve some kind of interdisciplinary base that can allow paradigm development and that produces scholarship which is credible to industry and government.

Conclusion

We have argued that action is the primary form of communication in the domain of cyber-security. Actors act, and seek legitimation through communication later. Despite much focus on the utility of strategic narratives for actors to articulate their vision of the past, present and future of international security, our analysis of a range of cases indicates that narratives are post-hoc and reactive to decisive actions. A further, important emerging dynamic to note is the entry to the normative market of new entrepreneurs. These are private companies which have begun to recognize the power that their ownership of these technologies grants them, as well as an increasing awareness of the implications of that power. These actors are likely to engage in ways that do not fit the typologies that we have previously examined. The role of non-state actors in the attribution of attacks carried out by states is a particular facet of this emerging phenomenon. Companies such as FireEye (2018) make claims as to the source and motivations of an attack, often proposing terms or time frames different from those that state actors provide. Similarly, insurance companies such as Zurich American refuse to pay out to customers on the grounds of a cyber-attack being an act of war (McCarthy, 2019). The politics of attribution and non-attribution is an interesting development to keep track of.

We have proposed that researchers should address the aesthetics of cyber-security, the effectiveness of the persuasion or securitization of those seeking to frame the Internet as a space of war on elite and public opinion, and the need for a basic mapping of the global architecture of cyber-security. Together these agendas allow for research that can explain how state and non-state action is conditioned by norms, cultures, gendered assumptions and technical capabilities – all matters of communication. The tools of digital politics provide a tremendous opportunity to build compelling explanations of interactions and developments in the domain of cyber-security.

Notes

1. A notable exception was a Special Eurobarometer study into European attitudes to cyber-security in 2013 (European Commission, 2013).
2. With apologies to Paul Virilio (2003), for whom the invention of the train was the invention of the train crash (and so for all technologies).

References

BBC News (2005) '1984: Tory Cabinet in Brighton Bomb Blast', 12 October. Available at http://news.bbc.co.uk/onthisday/hi/dates/stories/october/12/newsid_2531000/2531583.stm/ (accessed 17 July 2019).

Blumenthal, S. (1982) *The Permanent Campaign*. New York: Simon & Schuster.

Broad, W.J., Markoff, J. and Sanger, D.E. (2011) 'Israeli Test on Worm Called Crucial in Iran Nuclear Delay', *New York Times*, 15 January. Available at: https://www.nytimes.com/2011/01/16/world/middleeast/16stuxnet.html/ (accessed 17 July 2019).

Cohn, C. (1987) 'Sex and Death in the Rational World of Defense Intellectuals', *Signs: Journal of Women in Culture and Society*, 12(4), pp. 687–718.

Dunn Cavelty, M. (2013) 'From Cyber-Bombs to Political Fallout: Threat Representations with an Impact in the Cyber-Security Discourse', *International Studies Review*, 15(1), pp. 105–22.

European Commission (2013) 'Cyber Security', *Special Eurobarometer 404*, November. Available at http://ec.europa.eu/commfrontoffice/publicopinion/archives/ebs/ebs_404_en.pdf/ (accessed 17 July 2019).

FireEye (2018) 'Mandiant M-Trends2018', Special Report. https://www.fireeye.com/content/dam/collateral/en/mtrends-2018.pdf.

Fitzpatrick, M. (2019) 'Artificial Intelligence and Nuclear Command and Control'. *Survival Editors Blog*, 26 April. Available at https://www.iiss.org/blogs/survival-blog/2019/04/artificial-intelligence-nuclear-strategic-stability/ (accessed 23 July 2019).

Her Majesty's Government (2016) *Investigatory Powers Act 2016*. Available at: http://www.legislation.gov.uk/ukpga/2016/25/contents/enacted/ (accessed 2 September 2019).

Hudson, V.M., Ballif-Spanvill, B., Caprioli, M. and Emmett, C.F. (2012) *Sex and World Peace*. New York: Columbia University Press.

Keith, S. (2005) 'Fear-mongering or Fact: The Construction of 'Cyber-Terrorism' in US, UK, and Canadian News Media', Paper presented at Safety and Security in a Networked World: Balancing Cyber-Rights and Responsibilities. Sponsored by the Oxford Internet Institute, University of Oxford, 8–10 September. Available at https://www.oii.ox.ac.uk/archive/downloads/research/cybersafety/papers/susan_keith.pdf/ (accessed 17 July 2019).

Krippendorff, K. (2017) 'Three Concepts to Retire'. *Annals of the International Communication Association*, 41(1), pp. 92–9.

McCarthy, K. (2019) 'Cyber-Insurance Shock: Zurich Refuses to Foot NotPetya Ransomware Clean-Up Bill – and Claims it's an "Act of War"', *Register*, 11 January. Available at https://www.theregister.co.uk/2019/01/11/notpetya_insurance_claim/ (accessed 2 September 2019).

Olmstead, K. and Smith, A. (2017) 'What the Public Knows about Cybersecurity', Pew Research Center. Available at https://www.pewresearch.org/internet/ (accessed 6 November 2019).

Romm, T. and Harwell, D. (2019) 'Microsoft Workers Call for Canceling Military Contract for Technology that Could Turn Warfare into a "Video Game"', *Washington Post*, 22 February. Available at https://www.washingtonpost.com/technology/2019/02/22/microsoft-workers-call-cancelling-military-contract-technology-that-could-turn-warfare-into-video-game/?utm_term=.12511b42ba3a/ (accessed 17 July 2019).

Shlapak, D.A. and Johnson, M. (2016) *Reinforcing Deterrence on NATO's Eastern Flank: Wargaming the Defense of the Baltics*. Santa Monica, CA: RAND Corporation. Available at https://www.rand.org/pubs/research_reports/RR1253.html/ (accessed 16 July 2019).

Steele, B.J. (2010) *Defacing Power: The Aesthetics of Insecurity in Global Politics*. Ann Arbor, MI: University of Michigan Press.

Tooze, A. (2018) *Crashed: How a Decade of Financial Crises Changed the World*. New York: Penguin.

Valeriano, B. and Maness, Ryan C. (2015) *Cyber War Versus Cyber Realities: Cyber Conflict in the International System*. New York: Oxford University Press.

Virilio, P. (2003) *Unknown Quantity*. London: Thames & Hudson.

PART V

Reshaping Democratic Processes and Discourse

14 Beyond toxicity in the online public sphere: understanding incivility in online political talk

Patrícia Rossini

Introduction

With the promise of a digital public sphere capable of reinvigorating public debate and political participation far from being fulfilled, research on online political talk has been increasingly concerned with the challenges that digital platforms impose on democratically constructive conversations, such as the presence of uncivil discourse and trolling. While these issues have been discussed for over three decades, the rise of uncivil mainstream politics led by the upsurge of populism around the world, along with polarization and the prevalence of partisan media in high-choice environments, has sparked new interest around the potential effects of uncivil discourse, online and offline.

Several scholars have suggested that incivility in online discussions can undermine their value, leading to a dismissal of these forms of political engagement. Rather than dismissing online political talk due to the presence of incivility, researchers have been calling for a more nuanced approach to allow for a better understanding of the set of behaviours that are routinely classified as uncivil.

Uncivil by design? Disputing the democratic value of online political talk

The 'democratic potential' of the Internet to reinvigorate the public sphere has been a topic of scholarly concern for over three decades. The expectation that digital technologies would help to advance democratic goals was largely based on its technical features – or affordances – such as enabling people to talk to one another beyond geographic boundaries, to participate in discussions without fear of social sanctions or constraints, and the possibility of horizontal communication between citizens and their elected representatives (Coleman and Blumler, 2009). The hopes for a renewed public sphere were rapidly frustrated by empirical studies focused on online political discussions, e-government and e-participatory initiatives and political campaigns. Research suggested that the quality, tone and heterogeneity of online discussions were far from the standards of deliberative communication (Davis, 2005), that government-led digital initiatives were more focused on improving the efficiency of services than in fostering active citizen participation (Chadwick, 2003), and that political campaigns were not interested in opening channels for voters to interact with politicians (Stromer-Galley, 2000).

Despite being unable to fulfil the idealized goals of a 'virtual public sphere', online political talk remains one of the core areas of inquiry in the field of political communication, largely motivated by the variety of opportunities to participate in formal and informal discussion environments, such as forums, bulletin boards, news websites and, more recently, social media (Rossini and Stromer-Galley, 2019). The interest around online political discussions is not unjustified: talking about politics is a central practice to democratic citizenship, one that enables the public to learn about issues of public concern, build and recognize collective identities, learn about others' views and form, rehearse and clarify political opinions (Moy & Gastil, 2006; Rossini and Stromer-Galley, 2019).

Much of this scholarship has been framed within deliberative theory, focusing on the extent to which online political talk exhibits a set of normative discursive traits, such as rational exchange of arguments, justified positions, interpersonal respect, diversity of perspectives, reciprocity and civility (Coleman and Moss, 2012; Santana, 2014; Stromer-Galley and Wichowski, 2011). Others have called for a more inclusive approach, arguing that informal conversations should not be expected to live up to demanding deliberative norms (Chadwick, 2009; Freelon, 2010). While scholars may disagree about what constitutes democratically desirable conversation, most have consistently flagged incivility as a toxic feature of online political talk (Coe et al., 2014; Hmielowski et al., 2014; O'Sullivan and Flanagin, 2003; Santana, 2014). As a result, online discussions are routinely dismissed as democratically irrelevant due to a pervasive presence of incivility.

Many have argued that the Internet is uncivil by design: the affordances that were championed as being democratizing, such as anonymity, which would allow for participants to debate as equals, also might facilitate personal attacks, vulgar language or rudeness (e.g., Santana, 2014). The possibility of talking to others beyond geographic borders also means that people engage with unknown others, reducing the risk of social constraints that would force participants to 'save face' (Papacharissi, 2004). And while content moderation practices are largely seen as a solution to mitigate toxicity online and can have positive effects on the quality of discourse (Stroud et al., 2014), research has found that uncivil expressions prevail even in moderated environments (Rossini, 2019).

In short, research on online incivility has been focusing on a set of affordances of certain online channels to explain the phenomenon, but this line of inquiry becomes more complex as political talk takes place in multiple platforms; which calls for more comparative research. For instance, the idea that anonymity facilitates toxicity is challenged by the prevalence of incivility on social media, where people have personal profiles and use real names. Likewise, in the age of social media, moderation practices become less transparent as social media companies, notably known for not taking 'ownership' over the content in their platforms, are in charge of deciding the boundaries of what is permissible online (Gillespie, 2018). As a result, assumptions about the role of platform affordances in facilitating uncivil expressions need more scrutiny, as the changes in the media landscape make it more challenging to isolate which affordances are at play in facilitating or constraining these forms of expression.

Beyond understanding the role of platform affordances, research on online incivility has mainly adopted a view that these behaviours are detrimental to the quality of political talk. In what follows, I argue that scholars should move beyond this simplistic approach to incivility, which is a consequence of the lack of consensus around how incivility is conceptualized and measured. Then, I outline a research agenda for the future that focuses on better understanding the role of antinormative behaviours in online political talk.

Uncivil for whom?

Incivility is a challenging concept, and most scholars would agree that there is no consensus around its definition (Jamieson et al., 2015). A common reference to this lack of clarity is that 'incivility lies in the eye of the beholder' (Herbst, 2010; Jamieson et al., 2015). Most approaches to incivility fall in two theoretical traditions: politeness and deliberative democracy. In the former, behaviours such as rudeness and vulgar rhetoric, interpersonal disrespect, profanities, personal and ad hominem attacks, are considered violations of politeness and, as such, uncivil (Laden, 2019). In the context of deliberation, incivility is seen as a lack of respect for democratic norms and cooperation, unwillingness to acknowledge and engage with diverse positions, and to negotiate and compromise to benefit the public good (Mutz and Reeves, 2005).

Studies of online incivility have mainly adopted the politeness approach, focusing on behaviours such as name-calling, vulgarity, profane language, stereotyping, interpersonal disrespect, and graphic representations of shouting (writing with all caps or using many exclamation points) as signals of uncivil discourse (Chen, 2017; Coe et al., 2014; Hutchens et al., 2014; O'Sullivan and Flanagin, 2003; Rowe, 2015; Santana, 2014). Despite the overlap in the types of behaviours classified as uncivil, given differences in how scholars have conceptualized incivility, the results of current research cannot be compared. However, most studies looking at the presence of incivility in a variety of online platforms, from forums and news websites to social media, find comparable instances of incivility – ranging from 20 to 40 per cent – suggesting that, regardless of how one defines and measures it, the behaviour can be consistently observed in online political talk (Chen, 2017; Coe et al., 2014; Papacharissi, 2004; Rossini, 2019; Rowe, 2015; Santana, 2014).

The focus on incivility as impoliteness, instead of in the context of democratic norms, can be explained by an emphasis on a methodological focus on content analyses of online conversations. The features that indicate impoliteness are, to some extent, easier to operationalize as contextual cues: most people can spot and consistently classify these behaviours in written text; while underlying normative values such as cooperation, respect and consideration for diverse opinions, and focus on the public good, are not necessarily signalled as textual cues in online discourse. To some extent, these behaviours can be better identified in formal instances of political discussion, such as parliamentary or congressional debates, when cooperation, willingness to compromise, respect towards diverging ideas and focus on the public good are necessary to solve disagreements in order for political processes to unfold.

Given the challenges around how incivility online is defined and measured, a concerning result is that scholars have been conflating behaviours that signal face-to-face impoliteness with those that are more problematic and consequential, such as hate speech, violent threats, offensive stereotyping, and so on. When these are equated to name-calling, vulgar language, and so on, scholars cannot accurately judge the extent to which the presence of incivility means that online political talk is toxic or harmful.

It becomes equally challenging to understand the potential effects of incivility, a stream of research that has also prioritized the politeness approach. Studies suggest that the presence of incivility in online comments can undermine trust, and the perception of quality and credibility of news media outlets (Meltzer, 2015; Prochazka et al., 2018). Incivility in comments can increase the perception of risk and polarize the audience, among those who already held negative views on the topic (Anderson et al., 2014), or affect citizens' perceptions of the persuasiveness of others' arguments (Chen and Ng, 2016).

Less is known about the extent to which online incivility affects individual behaviour. A study found that blog posts can motivate readers to participate in the comments section, but they may also increase polarization about the topic (Borah, 2014). Others have found that incivility around partisan topics in the news can increase the perceptions that out-group politicians are behaving uncivilly, and in turn can discourage readers from engaging with the news (Muddiman et al., 2017). In the context of political discussions, there is some evidence that those who more routinely participate in online political talk are more likely to perceive incivility as acceptable, as well as more likely to engage in those behaviours, suggesting that the most active participants of online discussions are also the ones who perceive incivility to come with the territory when discussing politics (Hmielowski et al., 2014). On social media, incivility may even be perceived as entertaining (Sydnor, 2018). Interestingly, experimental research suggests that participants assume that uncivil comments can make others angry or upset, but do not report being angry or upset themselves, suggesting a 'third-person effect' (Chen and Ng, 2017) that could also be interpreted as an indication that the actual effects of being exposed to incivility are overstated, as people tend to think others will be offended even if they are not.

In the same vein as the stream of research investigating incivility in the content of online political talk, studies on effects also suffer from the conceptual fuzziness highlighted earlier in the chapter, which in turn suggests that the reference to the results of the studies summarized above must consider the different ways in which the concept was operationalized and measured. As a result, despite a large body of research analysing online incivility, the inferences that can be made from these studies are inconclusive at best in helping to determine whether these behaviours are harmful – or harmless – to online political talk.

Beyond 'just incivility'

To address the challenges around the concept of incivility, and accepting that the term may have different meanings depending on cultural and contextual situations, scholars have started to push for a nuanced approach. Papacharissi (2004) has argued that reducing incivility to face-to-face impoliteness ignores 'the democratic merit of robust and heated discussion'. Her work proposed a distinction between impoliteness and incivility, with the latter being restricted to behaviours that can be seen as a threat to democratic values, such as inciting violence towards democratic institutions, using offensive stereo-typing towards people and groups, and other types of behaviours that deny other people their freedoms (Papacharissi, 2004). This perspective makes an important contribution, arguing that the dismissal of impolite conversation may also mean denying that heated debate can be an important democratic activity; particularly when the affordances of mediated conversation can facilitate rudeness by removing some of the social constraints that are present in face-to-face interactions.

Thinking about incivility beyond digital environments, Muddiman et al. (2017) differentiates between personal-level incivility, grounded on politeness norms and, as such, violated by personal attacks and interpersonal rudeness, and public-level incivility, 'violations of reciprocity norms and disrespect for opposing political ideas' (Muddiman et al., 2017: 3199). In my own work on online political talk, I offer a distinction between incivility and intolerance, with the former referring to impolite, rude, vulgar or profane expressions, including name-calling and personal attacks, and the latter defined as expressions that violate moral respect and undermine individual and collective identities based on personal, social, sexual, ethnical, religious or cultural characteristics, violent threats, and so forth (Rossini, 2019).

There are several reasons to justify a nuanced approach to incivility to better understand its role in digital media. First, studies based on the distinction between impoliteness and incivility (Papacharissi, 2004; Rowe, 2015), or incivility and intolerance (Rossini, 2019), have found that the types of antinor-mative expressions that occur more frequently tend to fall under rudeness and vulgarity, while expressions that signal threats to democratic norms or values are rarely found in mainstream online spaces such as social media, news websites or popular online forums. Thus, it may be that the generalized concern around the toxicity of political talk online is exaggerated: while conversations may be heated and rude, there is not enough evidence to suggest that they are harmful or detrimental to advancing important democratic goals, such as being exposed to, and engaging with, the other side (Rossini, 2019).

Moreover, research examining citizens' perceptions suggests that some forms of incivility are accepted: for instance, name-calling and vulgarity are seen as more uncivil than messages containing aspersions or employing a pejorative tone (Kenski et al., 2017), and attacks on another person's conduct or character are seen as more uncivil than attacks on political arguments or positions (Muddiman et al., 2017; Stryker et al., 2016). In short, there is enough evidence that citizens' perceptions of incivility are not always aligned with what researchers deem as uncivil. Thus, scholars need to move beyond a single idea of incivility, and towards measuring presence, effects and perceptions of different types of behaviour.

Looking ahead: a research agenda for online incivility

This chapter has provided an overview of the research in online incivility and identifies the lack of clarity in what has counted as uncivil in current research, as well as the broad range of behaviours under the umbrella of 'incivility' as the main shortcomings in this agenda. To conclude, I turn to an agenda for future research to explore three areas: content, perceptions and effects.

First, if we accept that the nature of online discourse facilitates rudeness and profanities, because of affordances that reduce contextual cues and risk of sanctions, we must also consider that some expressions flagged as incivility might be considered normal in online discussions. Thus, scholars need to disentangle behaviours that denote heated arguments, interpersonal disrespect or lack of politeness from those that are democratically harmful – such as harassment, hate speech, violent threats and discrimination – in order to be able to distinguish democratically relevant conversation from toxic discourse. Still in the realm of content, scholars must consider that incivility might serve different rhetorical purposes, and distinguish expressions used to attack others from rhetoric that aims at advancing or reinforcing arguments.

Second, research has started to uncover the extent to which behaviours deemed by scholars as uncivil are indeed perceived as such by people, and the findings suggest that the public does not always align with scholarly definitions (Kenski et al., 2017; Stryker et al., 2016). So far, this work has not focused specifically on expressions of incivility online, for which participants may have different levels of tolerance that can be affected by experiences and other contextual variables, such as the platform where a comment is made, or whether or not participants are anonymous. Future research should further investigate how different types of antinormative expressions are received and perceived, adopting a cross-platform approach to unveil how contextual variables may affect the reception of antinormative behaviour.

Third, little has been uncovered about the types of discourse that are systematically moderated by platforms, news websites and online forums. As a consequence, what we know about online incivility based on public content is affected by moderation practices that may have introduced biases which scholars cannot fully understand or explain. While the concern about how platforms devise policy and practise moderation is not new (Dutton, 1996; Gillespie, 2018), the use of algorithms to perform moderation at scale has made these practices increasingly less transparent to the public. More research is needed to unveil how these moderation practices shape the tone of the content that is publicly visible.

Finally, the research investigating the effects of incivility online has suggested that this type of discourse can affect people's perception of the credibility, trustworthiness and quality of news outlets and articles, and can increase polarization. However, these studies have mainly focused on a few – often contentious – topics. More research is needed to understand how being exposed to incivility can affect future behaviour, such as the willingness to participate in a discussion, as well as individual-level effects, for example being offended, or feeling ostracized or silenced. Importantly, the inclusion of a broad range of behaviours under the same concept also means that the work around the effects of incivility has failed to address the effects of toxic forms of expression, such as harassment, violent threats, discrimination or hate speech. While these behaviours are less pervasive in most online places, their consequences are potentially more harmful both to participants in a discussion and to bystanders. They are also more challenging to study quantitatively, as these types of expressions could be moderated. Moving forward, a focus on qualitative research is particularly relevant to uncover the detrimental effects of online abuse.

To summarize, while incivility has benefited from a growing body of scholarly attention, there are several areas for this research agenda to develop. We must be careful with normative arguments for civility, as these often come from opinion leaders, elites and those in the centre of the public sphere, and can be used to ostracize marginalized voices, rather than foster a more democratic dialogue. Future research needs to step away from using 'incivility' as an umbrella term that encompasses a wide variety of behaviours, in order to better understand how antinormative discourse takes place online and, in particular, to be able to critically distinguish vulgar, profane or heated online banter from democratically alarming behaviours that silence and discriminate against others' voices.

References

Anderson, A. A., Brossard, D., Scheufele, D. A., Xenos, M. A., and Ladwig, P. (2014). The "Nasty Effect:" Online Incivility and Risk Perceptions of Emerging Technologies: Crude comments and concern. *Journal of Computer-Mediated Communication*, *19*(3), 373–387. https://doi.org/10.1111/jcc4.12009

Borah, P. (2014). Does It Matter Where You Read the News Story? Interaction of Incivility and News Frames in the Political Blogosphere. *Communication Research*, *41*(6), 809–827. https://doi.org/10.1177/0093650212449353

Chadwick, A. (2003). Bringing E-Democracy Back In: Why it Matters for Future Research on E-Governance. *Social Science Computer Review*, *21*(4), 443–455. https://doi.org/10.1177/0894439303256372

Chadwick, A. (2009). Web 2.0: New challenges for the study of e-democracy in an era of informational exuberance. *ISJLP*, *5*, 9.

Chen, G. M. (2017). *Online Incivility and Public Debate: Nasty Talk* (1st ed. 2017 edition). Palgrave Macmillan.

Chen, G. M., and Ng, Y. M. M. (2016). Third-person perception of online comments: Civil ones persuade you more than me. *Computers in Human Behavior*, *55*(Part B), 736–742. https://doi.org/10.1016/j.chb.2015.10.014

Chen, G. M., and Ng, Y. M. M. (2017). Nasty online comments anger you more than me, but nice ones make me as happy as you. *Computers in Human Behavior*, *71*(Supplement C), 181–188. https://doi.org/10.1016/j.chb.2017.02.010

Coe, K., Kenski, K., and Rains, S. A. (2014). Online and Uncivil? Patterns and Determinants of Incivility in Newspaper Website Comments. *Journal of Communication*, *64*(4), 658–679. https://doi.org/10.1111/jcom.12104

Coleman, S., and Blumler, J. G. (2009). *The Internet and Democratic Citizenship: Theory, Practice and Policy* (1st edition). Cambridge University Press.

Coleman, S., and Moss, G. (2012). Under Construction: The Field of Online Deliberation Research. *Journal of Information Technology & Politics*, *9*(1), 1–15. https://doi.org/10.1080/19331681.2011.635957

Davis, R. (2005). *Politics Online: Blogs, Chatrooms, and Discussion Groups in American Democracy* (1st edition). Routledge.

Dutton, W. H. (1996). Network rules of order: Regulating speech in public electronic fora: *Media, Culture & Society*. https://doi.org/10.1177/016344396018002006

Freelon, D. G. (2010). Analyzing online political discussion using three models of democratic communication. *New Media & Society*, *12*(7), 1172–1190. https://doi.org/10.1177/1461444809357927

Gillespie, T. (2018). *Custodians of the Internet: Platforms, Content Moderation, and the Hidden Decisions That Shape Social Media*. Yale University Press.

Herbst, S. (2010). *Rude democracy: Civility and incivility in American politics*. Temple University Press.

Hmielowski, J. D., Hutchens, M. J., and Cicchirillo, V. J. (2014). Living in an age of online incivility: Examining the conditional indirect effects of online discussion on political flaming. *Information, Communication & Society, 17*(10), 1196–1211. https://doi.org/10.1080/1369118X.2014.899609

Hutchens, M. J., Cicchirillo, V. J., and Hmielowski, J. D. (2014). How could you think that?!?!: Understanding intentions to engage in political flaming. *New Media & Society,* 1461444814522947.

Jamieson, K. H., Volinsky, A., Weitz, I., and Kenski, K. (2015). The Political Uses and Abuses of Civility and Incivility. In K. Kenski and K. H. Jamieson (Eds.), *The Oxford Handbook of Political Communication* (1st ed.). Oxford University Press. http://www.oxfordhandbooks.com/view/10.1093/oxfordhb/9780199793471.001.0001/oxfordhb-9780199793471

Kenski, K., Coe, K., and Rains, S. A. (2017). Perceptions of Uncivil Discourse Online: An Examination of Types and Predictors. *Communication Research,* 0093650217699933.

Laden, A. S. (2019). Two Concepts of Civility. In R. Boatright, T. Shaffer, S. Sobieraj, & D. G. Young, *A Crisis of Civility? Contemporary Research on Civility, Incivility, and Political Discourse.* (pp. 9–30). Routledge.

Meltzer, K. (2015). Journalistic Concern about Uncivil Political Talk in Digital News Media: Responsibility, Credibility, and Academic Influence. *The International Journal of Press/Politics, 20*(1), 85–107. https://doi.org/10.1177/1940161214558748

Moy, P., and Gastil, J. (2006). Predicting Deliberative Conversation: The Impact of Discussion Networks, Media Use, and Political Cognitions. *Political Communication, 23*(4), 443–460. https://doi.org/10.1080/10584600600977003

Muddiman, A., Pond-Cobb, J., and Matson, J. E. (2017). Negativity Bias or Backlash: Interaction With Civil and Uncivil Online Political News Content. *Communication Research,* 0093650216685625.

Mutz, D., and Reeves, B. (2005). The New Videomalaise: Effects of Televised Incivility on Political Trust. *American Political Science Review, 99*(01). https://doi.org/10.1017/S0003055405051452

O'Sullivan, P. B., and Flanagin, A. J. (2003). Reconceptualizing 'flaming' and other problematic messages. *New Media & Society, 5*(1), 69–94. https://doi.org/10.1177/1461444803005001908

Papacharissi, Z. (2004). Democracy online: Civility, politeness, and the democratic potential of online political discussion groups. *New Media & Society, 6*(2), 259–283. https://doi.org/10.1177/1461444804041444

Prochazka, F., Weber, P., and Schweiger, W. (2018). Effects of Civility and Reasoning in User Comments on Perceived Journalistic Quality. *Journalism Studies, 19*(1), 62–78. https://doi.org/10.1080/1461670X.2016.1161497

Rossini, P. (2019). Disentangling Uncivil and Intolerant Discourse. In R. Boatright, T. Shaffer, S. Sobieraj, and D. G. Young, *A Crisis of Civility? Contemporary Research on Civility, Incivility, and Political Discourse.* (pp. 142–157). Routledge.

Rossini, P., and Stromer-Galley, J. (2019). Citizen Deliberation Online. In E. Suhay, B. Grofman, and A. Tresel (Eds.), *Oxford Handbook of Electoral Persuasion.* Oxford University Press.

Rowe, I. (2015). Civility 2.0: A comparative analysis of incivility in online political discussion. *Information, Communication & Society, 18*(2), 121–138. https://doi.org/10.1080/1369118X.2014.940365

Santana, A. D. (2014). Virtuous or Vitriolic: The effect of anonymity on civility in online newspaper reader comment boards. *Journalism Practice, 8*(1), 18–33. https://doi.org/10.1080/17512786.2013.813194

Stromer-Galley, J. (2000). On-line interaction and why candidates avoid it. *Journal of Communication*, *50*(4), 111–132. https://doi.org/10.1111/j.1460-2466.2000.tb02865.x

Stromer-Galley, J., and Wichowski, A. (2011). Political discussion online. In M. Consalvo & C. Ess, *The handbook of internet studies* (Vol. 11, pp. 168–187). Wiley-Blackwell.

Stroud, N. J., Scacco, J. M., Muddiman, A., and Curry, A. L. (2014). Changing Deliberative Norms on News Organizations' Facebook Sites. *Journal of Computer-Mediated Communication*, *20*, 188–203. https://doi.org/10.1111/jcc4.12104

Stryker, R., Conway, B. A., and Danielson, J. T. (2016). What is political incivility? *Communication Monographs*, *83*(4), 535–556. https://doi.org/10.1080/03637751.2016.1201207

Sydnor, E. (2018). Platforms for Incivility: Examining Perceptions Across Different Media Formats. *Political Communication*, *35*(1), 97–116. https://doi.org/10.1080/10584609.2017.1355857

15 Facebook as a third space? The challenge of building global community

Scott Wright

Introduction

This chapter sets out the concept of the 'third space', and the potential importance of such spaces for mitigating the many challenges that afflict online political communication. More specifically, it focuses on the potential for Facebook to form (or facilitate) a third space. This is important because, while not explicitly using the phrase 'third space', in a major strategy reset that set out his vision for Facebook, chief executive officer Mark Zuckerberg (2017) argued that Facebook is a form of global community and that, in spite of various challenges (for example, disinformation, false news), it should continue to be the place where global community is built online. As outlined in more detail below, the vision that Zuckerberg sets out echoes third space (and to an extent, 'third place') theorizing in various ways. To borrow from Howard Rheingold's seminal work on online communities and his reflections on whether the WELL online community was a third place:

> [Facebook] might not be the same kind of place that Oldenburg had in mind, but so many of his descriptions of third places could also describe [Facebook]. Perhaps [Facebook] is one of the informal public places where people can rebuild the aspects of community that were lost when the malt shop became a mall. (Rheingold, 1993, 23–4)

Both Zuckerberg and Rheingold discuss the challenges of building community online, but ultimately argue that the challenges can be overcome and community can be built. Zuckerberg, in particular, argues that we should trust Facebook (and him) as the place to build the social infrastructure of global online community. Given the potential civic and political importance of third spaces, the significance placed by Zuckerberg on the potential for Facebook to facilitate a form of third space, and the hugely important role that Facebook plays in many societies, this chapter argues that the extent to which Facebook actually comprises a third space is an important research agenda for digital politics researchers going forward. It will be argued, though, that this is a challenging and complex research agenda to prosecute. While there have been a small but growing number of studies that have appropriated either third place (Oldenburg, 1999) or third space theory (Wright, 2012a, 2012b) to study political talk (e.g., Graham and Wright, 2014; Kligler-Vilenchik, 2015) and participation (e.g., Graham et al., 2015, 2016), such research is limited, and inhibited by several challenges. This chapter outlines some of these challenges and potential avenues to take research forward.

Third space

In its simplest form, a third space is a formally non-political online space where political talk emerges. The concept of third space builds on a critique of Ray Oldenburg's (1999) concept of the third place. A third place, according to Oldenburg, 'is a generic designation for a great variety of public spaces that host the regular, voluntary, informal, and happily anticipated gatherings of individuals', and is a 'core setting of informal public life' (ibid.: 16), including pubs and cafes and community hangouts. Not all pubs and cafes are third places though. Oldenburg highlights a series of characteristics that are central to the construction of a third place. These include that they must be on neutral and accessible ground; act as a social leveller; with conversation as the main activity; the mood must be playful; and a group of regulars must help to set the tone of debate (ibid.: 22–42).

Oldenburg believes that third places perform a crucial role in the development of societies and communities, helping to strengthen citizenship, and thus are 'central to the political processes of a democracy' (Oldenburg, 1999: 67). Moving beyond Oldenburg for a moment, the kinds of everyday online political talk that might occur in a third place are considered the lifeblood of the public sphere (Habermas, 1974: 49), and crucial to democratic citizenship (Dahlgren, 2006: 282) and deliberative democracy because 'through everyday political talk, citizens construct their identities, achieve mutual understanding, produce public reason, form considered opinions' (Kim and Kim, 2008: 51), facilitating political knowledge, engagement, opinion change (Price and Cappella, 2002; Huckfeldt et al., 2004) and political actions (Graham et al., 2015, 2016). Importantly, third spaces are often the domain of 'ordinary' citizens and hard-to-reach voters (or potential voters), rather than the political animals that tend to dominate political forums and hashtags.

Oldenburg, however, was deeply sceptical of the role of the Internet, suggesting that it harmed the kinds of informal political socialization that he cherished. Oldenburg (1999: 204) argued that, '[a]n efficient home-delivery media system, in contrast, tends to make shut-ins of otherwise healthy individuals', and so 'atomized the citizenry that the term "society" may no longer be appropriate'. Nevertheless, early online community scholars argued that online communities bore many similarities to Oldenburg's third places (Rheingold, 1993: 10; Schuler, 1996).

Early empirical research suggested that online communities might be fruitful sites for political talk, and some echoed Oldenburg in their language. Wojcieszak and Mutz (2009: 45) found that 53 per cent of Americans engaged in political talk in 'non-political' online communities, and that this was more likely to cut across political lines. Steinkuehler and Williams's (2006) study of online game communities found that they were 'new (albeit virtual) "third places" for informal sociability', while Soukup (2006) delved more into the theory, and more recently McArthur and White (2016) employed Oldenburg's third place in the context of Twitter debates.

In previous work, I have retheorized the concept of the third place as a third space (Wright, 2012a, 2012b). This approach accounts for some of the differences between online and face-to-face communities, and provides a theoretically informed route forward for assessing everyday online political talk. While there is insufficient space here to provide a full account of third space (see Wright, 2012b in particular for a more detailed account), these differences include affording value to geographically distant, issue-based communities and not just physically proximate ones; and barriers such as requiring people to login, the use of real name identifiers, and the use of moderators and community guidelines to manage debates are things that either already exist in some form in the 'real world' or are necessary because online communication is so different.

The potential value of third spaces

It is believed that third spaces may help to avoid or limit many of the pitfalls that inflict online political debate. First, there are significant concerns about political polarization online (Smith et al., 2014). Polarization harms deliberation as people can avoid cross-cutting political talk, and may lead people to become more trenchant and less tolerant as they consume information that reinforces rather than challenges their beliefs (Sunstein, 2001; Huckfeldt et al., 2004: 11; Delli Carpini et al., 2004). As third spaces are not politically focused – people go there to discuss hobbies (for example, fishing, crafting, gardening, sport) or receive support (for example, parenting, financial advice, education) – they are much less likely to be politically polarized. As Graham and Harju (2011: 29) argue, 'fragmentation theory makes little sense once we move beyond the politically oriented communicative landscape' (Graham and Harju, 2011: 29). This is supported by Wojcieszak and Mutz's (2009: 45) United States (US)-focused survey, which suggests that talk in everyday online spaces features more cross-cutting talk than political spaces. However, it is worth noting that the topic of the community might affect the extent of political polarization (Yan et al., 2018). For example, we might expect hunting or sport shooting communities to lean to the right, and recycling communities to lean to the left (though this remains an empirical question in need of further research).

A study of political talk about the 2016 Australian federal election in an online parenting community found that debates did cut across political lines, with people from across the political spectrum present, participating and disagreeing (Wright et al., 2018). Political difference was handled in a supportive manner. For example, when one mother expressed support for a far-right politician, the other parents supported her personally (Wright et al., 2018). It is perhaps worth noting that debates were also geographically cross-cutting; that is, they were third spaces rather than geographically tight third places. Yan et al.'s (2018) analysis of online cricket debates found disagreement rates at above 50 per cent across all topics.

Second, many people choose to avoid political talk, whether because they dislike conflict or because they do not feel informed enough to participate (Eliasoph, 1998; Conover and Searing, 2005; Mutz, 2006). In extreme cases, people may leave groups or 'unfriend', 'mute' or 'block' people (see, e.g., John and Dvir-Gvirsman, 2015). Online, around 80 per cent of people do not visit political spaces (Horrigan et al., 2001), but prefer to 'hang out' in social online communities (Ridings and Gefen, 2004). As political talk can emerge anywhere in such third spaces and is often unexpected, it is, I contend, much harder to avoid. Following Lev-On and Manin (2009), there can be 'happy accidents', or what Brundidge (2011: 687) describes as inadvertent exposure.

Research has found extensive informal and formal political talk in a wide range of third spaces. Political talk emerges across forums, and is just as likely to emerge during the course of a discussion thread as it is for a thread to start about politics. A study of a financial help forum (Graham and Wright, 2014) found that 32 per cent of threads contained political talk, and that it emerged in just under half of all of the sub-forums, from discount codes to pet care. This mirrors Graham and Harju's (2011) analysis of the *Wife Swap* TV programme forum, Graham's (2010) analysis of the *Big Brother* TV programme forum, Van Zoonen's (2007) work on film communities, and Yan et al.'s (2018) study of cricket communities. Alongside the finding that political disagreements can (at least sometimes) be handled with civility, the research found that comments that degrade another participant (2 per cent) or attempt to curb their commenting (1 per cent) were minimal, supporting the contention that the different forms of avoidance are less likely (Graham and Wright, 2014), though this varies from study to study.

Third, there are extensive concerns about incivility, trolling and abuse in online political debate, with people speaking 'past' each other rather than listening and engaging in debate (Wilhelm, 2000). As indicated above, incivility has been found to be quite rare, at least in terms of curbing and degrading attacks (Graham and Wright, 2014). This finding is supported by research on election debates in parenting forums (Wright et al., 2018), and a broader study of political talk in the same forum (Wright, 2018) assessing differences between 'super-posters' and the broader community. This found that incivility was rare across the board, and that super-posters were slightly more likely to degrade (3.4 per cent, 2.3 per cent), curb (1.6 per cent, 0.9 per cent) and flame (1.5 per cent, 0.7 per cent). Following Oldenburg, it is questionable whether Habermas-inspired notions of civility make sense in the context of everyday online debates, which can be heavily focused on chat, playfulness and the use of rudeness as a bonding mechanism (Svensson, 2015; Rowe, 2017). Yan et al. (2018: 1583) found that when an opinion was expressed in their cricket forums (75.1 per cent of posts), only 24.8 per cent gave a reason to support their opinion, and when they disagreed they were more likely to be uncivil ($t = 3.44$, $p < 0.001$), concluding that flaming was 'likely a hallmark of cross-cutting political discourse'.

Fourth, in an apparent era of declining political participation, third spaces have the potential to be participation incubators; through everyday discourse and social encouragement, people may become more political and provide spaces for self-actualizing citizenship (Kligler-Vilenchik, 2015). This analysis builds on a deeper history of how political talk (Huckfeldt and Sprague, 1995), and online political talk in particular (Gil de Zuniga and Valenzuela, 2011), support political participation. In Dahlgren's (2006) terms, they encourage proto-political behaviours that foreground a turn to political action. Research has found that acts and calls for political participation are widespread in third spaces. Graham et al. (2015, 2016) found extensive evidence of (acts of, or calls for) both manifest political participation (for example, voting, signing petitions) and latent political participation (for example, changing clothing, 'buycots') in two-thirds of spaces that had a strong social and help function (financial and parenting), but less in a forum devoted to the discussion of television (which also had an explicitly political sub-forum). Interestingly, personal actions (for example, navigating the welfare system) can turn into broader public actions, suggestive of the incubator function.

This section has explained the theoretical and practical importance of third spaces, highlighting four principal features. In doing this, it has also covered a range of literature that has sought to assess political talk and its implications, and the nascent literature that has assessed online political talk in (apparent) third spaces. The next section discusses the important, but often derided, Building Global Community letter.

Building global community: defining Facebook as a third space?

The Building Global Community letter was the first significant update to Facebook's mission since the 2012 mission statement published at the time of the Facebook initial public offering (IPO). The original argued that Facebook's 'social mission [was] to make the world more open and connected'. Facebook, and its brief precursor, FaceMash, are ego-centred networks. That is, Facebook was focused on the individual profile and their friendship networks, with the News Feed subsequently surfacing information using an algorithm that included information such as people's previous behaviour. Groups, in contrast, bring people together focused not on their strong and weak ties, but on their social (and other) interests.

Zuckerberg's thinking about Facebook as a community has oscillated. The original goal, according to Zuckerberg (cited in Garber, 2012), 'wasn't to make a huge community site, it was to make something where you could type in someone's name and find out a bunch of information about them . . . the goal that we went into it with wasn't to make an online community but sort of like a mirror for the real community that existed in real life' (though successful examples whereby Facebook promoted on- and offline community were celebrated; Zuckerberg, 2009; see Hoffman et al., 2016).

The rationale that underpinned the penning of Building Global Community is an important question. While it is often presented as a response to different crises, Zuckerberg claims that the genesis of the letter predates the 2016 US Federal election and was sparked by a realization that Facebook had to change in response to the blocking of 'napalm girl' by Facebook moderators (Manjoo, 2017). It is, however, a significant jump from this incident to what is presented in Building Global Community, and the timing of its publication is clearly significant.

The way in which Zuckerberg presents Facebook in Building Global Community has been a subject of intense, often critical debate. People have presented the document as akin to a political manifesto (Rider and Murakami Wood, 2019); a smokescreen for a large corporation whose goal is to maximize profit rather than social good; and a public relations exercise that seeks to rebrand both Zuckerberg (who has been presented as a cut-throat business person in *The Network*) and Facebook, which had suffered a series of crises from 'false news' and misinformation to data privacy (Dreyfuss, 2017; *Guardian*, 2017; Oremus, 2017; Heer, 2017). The document is controversial, not least because it argues for global connectivity at a time when there has been a shift to more isolationism (Isaac, 2017).

While some of the criticisms may have merit, it remains an important document in need of deeper scrutiny because the discourses used to describe Facebook help us to understand the company and its future direction (Hoffman et al., 2016). This section will focus on how Zuckerberg presents Facebook, and particularly the ways in which this echoes third space theorizing. Zuckerberg (2017) states that the higher-level question is not profit, but whether we are 'building the world we all want', as this is the 'most important question of all'. He argues, 'Facebook stands for bringing us closer together and building a global community.' Zuckerberg presents Facebook's contribution to this across five streams in which Facebook can help to build supportive, safe, informed, civically engaged and inclusive communities.

First, Zuckerberg (2017) echoes Oldenburg (1999) and Putnam (2000), arguing that there has been a 'striking decline in the important social infrastructure of local communities' such as churches, sporting groups, and unions. While Oldenburg and Putnam mainly see potential for damage by a rise in online communities – exacerbating what Oldenburg sees as the 'problem of place', Zuckerberg echoes third-space theory, which argues that we should not 'privilege place over issue-based (and related) forums and communities' (Wright, 2012b: 11), and that online community can support offline interaction, particularly where there are geographic as well as social or issue-based ties.

For Zuckerberg (2017; see also Zuckerberg, 2009): 'Online communities are a bright spot, and we can strengthen existing physical communities by helping people come together online as well as offline.' To this end, Zuckerberg wants to improve how Facebook promotes what he calls 'meaningful groups', in particular, and the tools that enable community leaders to create them. For Facebook, this seems to mean a shift in focus from the News Feed to different forms of community, or in Facebook parlance, Groups. The shift to emphasize Groups is in some ways arguably a return to the online communities of the 1990s–2000s, and while Groups are not a new feature for Facebook, the focus on Groups is a step change in need of research.

Second, Zuckerberg outlines the importance of safe communities, discussing both Facebook's role in responding to events such as natural disasters, terrorist attacks or missing children (for example, Safety Check) and, more pressing in the context of third spaces, how Facebook polices its community and attempts to stop 'bad actors'. There is no direct correlation here with third space theory to date; it really cuts across the different structural and social characteristics. Communication in third spaces may not be the rational-critical communication envisaged by deliberative democrats. It often involves humour or banter and may even, as Oldenburg notes, be designed to offend amongst close friends. How to moderate such communication is a central question for Facebook, third spaces and the regulation of platforms in general: where are the acceptable boundaries of communication, and what does this mean for questions of power in a democratic society (Gillespie, 2018; Suzor, 2019)? As defined in Building Global Community, Zuckerberg (2017) presents a more libertarian approach that 'will focus less on banning misinformation, and more on surfacing additional perspectives and information, including that fact checkers dispute an item's accuracy'.

The power of moderation also speaks to issues of neutrality, inclusivity and access, and the economic logics that shape moderation processes. While Oldenburg emphasized informal norms of conversation and 'the regulars' in third places as regulatory devices, even with online 'super-participants' who help to maintain norms and can act as moderators (Graham and Wright, 2014; Wright, 2018), the more fluid nature of online interaction means that moderation is important to debate quality (Wright, 2006; Wright and Street, 2007); though again, moderation in general, and Facebook's approach in particular, remains another important stream for digital politics researchers.

Third, as part of creating an informed community, Zuckerberg wants to give 'everyone a voice [which] has historically been a very positive force for public discourse because it increases the diversity of ideas shared'. Here Zuckerberg addresses the challenge of filter bubbles, online polarization and fake news. As discussed above, at least on the non-political defined parts of Facebook, third space theory would expect polarization to be limited and the diversity of opinions to inhibit filter bubbles (Wright, 2012b; Bruns, 2019). Zuckerberg expounds on this at length:

> Research suggests the best solutions for improving discourse may come from getting to know each other as whole people instead of just opinions – something Facebook may be uniquely suited to do. If we connect with people about what we have in common – sports teams, TV shows, interests – it is easier to have dialogue about what we disagree on. When we do this well, we give billions of people the ability to share new perspectives while mitigating the unwanted effects that come with any new medium.

This paragraph strongly echoes the arguments that underpin third space (and third place) theory and the potential benefits therein. It is not clear what research Zuckerberg draws on here, though. Zuckerberg again echoes the logic of third spaces, in which 'conversation is the main activity' (Oldenburg, 1999: 26), when stating that: 'I want to emphasize that the vast majority of conversations on Facebook are social, not ideological. They're friends sharing jokes and families staying in touch across cities.'

Fourth, Zuckerberg focuses on civic engagement, arguing that Facebook can both encourage people to engage in existing processes and, more radically, experiment with new forms of community engagement. Research has found that third spaces generate a wide range of political actions (Graham et al., 2015, 2016), but research is needed into both how this happens organically on Facebook, and how Facebook's promised new tools support civic engagement.

Finally, Zuckerberg turns to look at inclusive community, noting that Facebook governance has failed at times; a topic that intersects his different themes. Zuckerberg blames these challenges on several issues, including the scale and complexity of communication and related moderation processes. When he argues that 'our community is evolving from its origin connecting us with family and friends to now becoming a source of news and public discourse as well', this seems to be more about how Facebook sees itself; it has long been a source of news and discourse, even if focused on people's 'walls' (personal pages).

Zuckerberg emphasizes the need for local governance here, though whether Facebook is setting these standards or reacting to nation-state level laws, and how effective this is, can still be considered an emerging research field. But Building Global Community gives us useful markers for how Zuckerberg himself sees Facebook. This includes giving 'everyone in the community options for how they would like to set the content policy for themselves', which sits underneath the broader community standards. While this makes intuitive sense in such a vast community, there is the potential here for people to 'avoid' disagreement, and it may inhibit the construction of a third space. This sits alongside algorithmic narrowcasting, which means that people tend to see material the algorithm believes they are interested in. In other words, there is not one 'common' group experience, and how this impacts upon the construction of third space on Facebook, and the power the algorithm gives Facebook in the construction of community (for example, what groups are surfaced to what individuals), is a particularly important question (Couldry, 2015).

While there is insufficient space here to discuss this in detail, I must briefly also mention what is downplayed or ignored in Building Global Community. Perhaps unsurprisingly, the power that Facebook wields in how it builds what it terms 'social infrastructure', and the economic incentives that underpin this, are largely ignored. The political economy of Facebook is an important question. The increasingly dominant position of Facebook raises significant concerns for its function as a third space. While third places and third spaces may be commercial, one suspects that Oldenburg (and perhaps Rheingold) might see Facebook as akin to the mall taking over and harming the diversity that existed online from the wide range of independently owned and operated forums and listserves.

The challenge of analysing Facebook as a third space

While there is a long track record of research into online communities, whether or not Facebook is a 'bright spot' for online community remains an important and open question. In this section, some of the challenges of assessing everyday online political talk in the 'non-political' parts of Facebook that might constitute a third space are discussed. Capturing and assessing everyday online political talk on Facebook is hard for three principal reasons.

First, while such talk is huge in volume on Facebook, it is diluted within the whole pool of Facebook activity: it is like searching for needles in a haystack (Graham, 2008). Assessing political talk on the Facebook page of a politician is (or was) relatively simple: we can start with the assumption that all of the content there is political. Complicating matters further, what should constitute politics or, more broadly, the political, is hotly debated. And on top of this, the concept of talk – as opposed to, say, deliberation or discourse – is contested.

Turning such concepts into operationalizable definitions for content analysis is tricky. Qualitative research, such as ethnography and discourse analysis, also holds significant promise (Kligler-Vilenchik, 2015; Rowe, 2017).

Second, non-politically defined groups are often hard to reach: they are closed-access, private or semi-private Facebook Groups, and assessing the political talk therein – even on public Facebook groups – raises significant ethical questions, such as those around informed consent. By contrast, the oft-studied politicians' Facebook pages are easier research targets because they are public figures and normally open access.

This builds into a related third issue: after the Cambridge Analytica scandal, which involved a major privacy breach linked to a researcher, Facebook has restricted or closed many of its application programming interfaces (APIs) that have provided access to data for researchers (Granville, 2018). While technically it is possible to scrape content from the front-end of public Facebook pages (subject to ethical, copyright and terms-of-service restrictions), Facebook does not display all of the content, and algorithms shape what is displayed. Social Science One also provides some access to data for researchers, but the 'API graveyard' creates a very different world that we are only just starting to explore (Digital Methods, 2019).

Conclusion

The vision for Facebook embedded in Building Global Community is worthy of study. Empirically, whether or not Facebook delivers on this shift to global community is in need of research, as is the more specific question of whether (parts of) Facebook constitute a third space. This chapter has outlined several avenues for new research, as well as the challenges of prosecuting this research agenda. How Facebook collaborates with researchers – via Social Science One or some other mechanism – is particularly important. While there is a significant amount of doom and gloom, there are seeds of positivity too. Given Facebook's focus on Building Global Community – and how this echoes third space logic – this is, I believe, one area where fruitful collaboration is possible and important.

References

Brundidge, J. (2011) 'Encountering "Difference" in the Contemporary Public Sphere: The Contribution of the Internet to the Heterogeneity of Political Discussion Networks', *Journal of Communication*, 60(4): 680–700.

Bruns, A. (2019) *Are Filter Bubbles Real?* Cambridge: Polity.

Conover, P. and Searing, D. (2005) 'Studying "Everyday Political Talk" in the Deliberative System', *Acta Politica*, 40, pp. 269–83.

Couldry, N. (2015) 'The Myth of "Us": Digital Networks, Political Change and the Production of Collectivity', *Information, Communication and Society*, 18(6), pp. 608–26.

Dahlgren, P. (2006) 'Doing Citizenship: The Cultural Origins of Civic Agency in the Public Sphere', *European Journal of Cultural Studies*, 9(3), pp. 267–86.

Delli Carpini, M., Cook, F.L. and Jacobs, L.R. (2004) 'Public Deliberation, Discursive Participation, and Citizen Engagement: A Review of the Empirical Literature', *Annual Review of Political Science*, 7, pp. 315–44.

Digital Methods (2019) 'Post-API Research? On the Contemporary Study of Social Media Data'. Available at https://wiki.digitalmethods.net/Dmi/WinterSchool2020/ (accessed 16 October 2019).

Dreyfuss, E. (2017) 'The Community Zuck Longs to Build Remains a Distant Dream', *Wired*. Available at: https://www.wired.com/2017/05/community-zuck-longs-build-remains-distant-dream/ (accessed 16 October 2019).

Eliasoph, N. (1998) *Avoiding Politics*. Cambridge: Cambridge University Press.

Garber, M. (2012) 'The Ballad of Mark Zuckerberg', *Atlantic*, 1 February. https://www.theatlantic.com/technology/archive/2012/02/the-ballad-of-mark-zuckerberg/252374/.

Gil de Zuniga, H. and Valenzuela, S. (2011) 'The Mediating Path to a Stronger Citizenship: Online and Offline Networks, Weak Ties and Civic Engagement', *Communication Research*, 38, pp. 397–421.

Gillespie, T. (2018) *Custodians of the Internet: Platforms, Content Moderation, and the Hidden Decisions That Shape Social Media*. New Haven, CT: Yale University Press.

Graham, T. (2008) 'Needles in a Haystack: A New Approach for Identifying and Assessing Political Talk in Nonpolitical Online Spaces', *Javnost – The Public*, 15(2), pp. 17–36.

Graham, T. (2010) 'Talking Politics Online within Spaces of Popular Culture: The Case of the Big Brother Forum', *Javnost – The Public*, 17(4), pp. 25–42.

Graham, T. and Harju, A. (2011) 'Reality TV as a Trigger of Everyday Political Talk in the Net-Based Public Sphere', *European Journal of Communication*, 26(1), pp. 18–32.

Graham, T., Jackson, D. and Wright, S. (2015) 'From Everyday Conversation to Political Action: Talking Austerity in Online "third spaces"', *European Journal of Communication*, 30(6), pp. 648–65.

Graham, T., Jackson, D. and Wright, S. (2016) '"We Need to Get Together and Make Ourselves Heard": Everyday Online Spaces as Incubators of Political Action', *Information, Communication and Society*, 19(10), pp. 1373–89.

Graham, T. and Wright, S. (2014) 'Discursive Equality and Everyday Political Talk: The Impact of Super-Participants', *Journal of Computer-Mediated Communication*, 19(3), pp. 625–42.

Granville, K. (2018) 'Facebook and Cambridge Need to Analytica: What You Know as Fallout Widens', *New York Times*. Available at https://www.nytimes.com/2018/03/

19/technology/facebook-cambridge-analytica-explained.html/ (accessed 16 October 2019).

Guardian (2017) 'The Facebook Manifesto: Mark Zuckerberg's Letter to the World Looks a Lot like Politics'. Available at https://www.theguardian.com/technology/shortcuts/2017/feb/17/facebook-manifesto-mark-zuckerberg-letter-world-politics/ (accessed 16 October 2019).

Habermas, J. (1974) Theory and Practice. London: Heinemann.

Heer, J. (2017) #Facebook's Promise of Community Is a Lie', New Republic. Available at https://newrepublic.com/article/145213/facebooks-promise-community-lie/ (accessed 16 October 2019).

Hoffman, A.L., Proferes, N. and Zimmer, M. (2016) '"Making the World More Open and Connected": Mark Zuckerberg and the Discursive Construction of Facebook and Its Users', New Media and Society, 20(1), pp. 199–218.

Horrigan, J.B., Rainie, L. and Fox, S. (2001) Online Communities: Networks that Nurture Long-Distance Relationships and Local Ties. Washington, DC: Pew Internet and American Life Project.

Huckfeldt, R., Mendez, J.M. and Osborn, T.L. (2004) 'Disagreement, Ambivalence, and Engagement: The Political Consequences of Heterogeneous Networks', Political Psychology, 25(1), pp. 65–95.

Huckfeldt, R. and Sprague, J. (1995) Citizens, Politics, and Social Communication: Information and Influence in an Election Campaign. Cambridge: Cambridge University Press.

Isaac, M. (2017) 'Facebook's Zuckerberg, Bucking Tide, Takes Public Stand Against Isolationism', New York Times. Available at https://www.nytimes.com/2017/02/16/technology/facebook-mark-zuckerberg-mission-statement.html.

John, N.A. and Dvir-Gvirsman, S. (2015) '"I Don't Like You Any More": Facebook Unfriending by Israelis During the Israel–Gaza Conflict of 2014', Journal of Communication, 65(6), pp. 953–74.

Kim, J. and Kim, E.J. (2008) 'Theorizing Dialogic Deliberation: Everyday Political Talk as Communicative Action and Dialogue', Communication Theory, 18(1), pp. 51–70.

Kligler-Vilenchik, N. (2015) 'From Wizards and House-Elves to Real-World Issues: Political Talk in Fan Spaces', International Journal of Communication, 9, pp. 2027–46.

Lev-On, A. and Manin, B. (2009) 'Happy Accidents: Deliberation and Online Exposure to Opposing Views'. In T. Davies and S.P. Gangadharan (eds), Online Deliberation: Design, Research, and Practice. Palo Alto, CA: CSLI, pp. 105–22.

Manjoo, F. (2017) 'Can Facebook Fix Its Own Worst Bug?', New York Times. Available at https://www.nytimes.com/2017/04/25/magazine/can-facebook-fix-its-own-worst-bug.html/ (accessed 16 October 2019).

McArthur, J.A. and White, A.F. (2016) 'Twitter Chats as Third Places: Conceptualizing a Digital Gathering Site', Social Media+Society, 2(3), pp. 1–9.

Mutz, D. (2006) Hearing the Other Side: Deliberative Versus Participatory Democracy. Cambridge: Cambridge University Press.

Oldenburg, R. (1999) The Great Good Place: Cafes, Coffee Shops, Bookstores, Bars, Hair Salons, and Other Hangouts at the Heart of a Community. New York: Marlowe & Company.

Oremus, W. (2017) 'Facebook's New 'Manifesto' Is Political. Mark Zuckerberg Just Won't Admit It', Slate. Available at https://slate.com/technology/2017/02/the-problem-with-mark-zuckerbergs-new-facebook-manifesto-it-isnt-political-enough.html/ (accessed 16 October 2019).

Price, V. and Cappella, J.N. (2002) 'Online Deliberation and Its Influence: The Electronic Dialogue Project in Campaign 2000', *IT and Society*, 1(1), pp. 303–29.

Putnam, R.D. (2000) *Bowling Alone: The Collapse and Revival of American Community*. New York: Simon & Schuster.

Rheingold, H. (1993) *The Virtual Community: Homesteading on the Electronic Frontier*. Reading, MA: Addison-Wesley.

Rider, K and Murakami Wood, D. (2019) 'Condemned to Connection? Network Communitarianism in Mark Zuckerberg's "Facebook Manifesto"', *New Media and Society*, 21(3), pp. 639–54.

Ridings, C.M. and Gefen, D. (2004) 'Virtual Community Attraction: Why People Hang Out Online', *Journal of Computer-Mediated Communication*, 10(1), pp. 1–10.

Rowe, P. (2017) 'The Everyday Politics of Parenting: A Case Study of MamaBake', *Journal of Information Technology and Politics*, 15(1), pp. 34–49.

Schuler, D. (1996) *New Community Networks*. New York: ACM Press.

Smith, M.A., Rainie, L., Shneiderman, B. and Himelboim, I. (2014) 'Mapping Twitter Topic Networks: From Polarized Crowds to Community Clusters', Pew Research Internet Project. Available at www.pewinternet.org/2014/02/20/mapping-twitter -topic-networks-from-polarized-crowds-tocommunity-clusters/ (accessed 16 October 2019).

Soukup, C. (2006) 'Computer-Mediated Communication as a Virtual Third Place: Building Oldenburg's Great Good Places on the World Wide Web', *New Media and Society*, 8, pp. 421–40.

Steinkuehler, C.A. and Williams, D. (2006) '"Where Everybody Knows Your (Screen) Name": Online Games as "Third Places"', *Journal of Computer-Mediated Communication*, 11, pp. 885–909.

Sunstein, C. (2001) *Republic.com*. Princeton, NJ: Princeton University Press.

Suzor, N.P. (2019) *Lawless: The Secret Rules that Govern Our Digital Lives*. Cambridge: Cambridge University Press.

Svensson, J. (2015) 'Participation as a Pastime: Political Discussion in a Queer Community Online', *Javnost*, 22(3), pp. 283–97.

Van Zoonen, L. (2007) 'Audience Reactions to Hollywood Politics', *Media, Culture and Society*, 29(4), pp. 531–47.

Wilhelm, A.G. (2000) *Democracy in the Digital Age: Challenges to Political Life in Cyberspace*. London: Routledge.

Wojcieszak, M. and Mutz, D. (2009) 'Online Groups and Political Discourse: Do Online Discussion Spaces Facilitate Exposure to Political Disagreement?', *Journal of Communication*, 59(1), pp. 40–56.

Wright, S. (2006) 'Government-Run Online Discussion Fora: Moderation, Censorship and the Shadow of Control', *British Journal of Politics and International Relations*, 8(4), pp. 550–68.

Wright, S. (2012a) 'Politics as Usual? Revolution, Normalization and a New Agenda for Online Deliberation', *New Media and Society*, 14(2), pp. 244–61.

Wright, S. (2012b) 'From "Third Place" to "Third Space": Everyday Political Talk in Non-Political Online Spaces', *Javnost – The Public*, 19(3), pp. 5–20.

Wright, S. (2018) 'The Impact of "Super-Participants" on Everyday Political Talk', *Journal of Language and Politics*, 17(2), pp. 155–72.

Wright, S., Lukamto, W. and Trott, V. (2018) 'The 2016 Australian Election Online: Debate, Support, Community'. In A. Gauja, P. Chen, J. Curtin and J. Pietsch (eds), *Double Disillusion: The 2016 Australian Federal Election*. Canberra: ANU Press.

Wright, S. and Street, J. (2007) 'Democracy, Deliberation and Design: The Case of Online Discussion Forums', *New Media and Society*, 9(5), pp. 849–69.

Yan, W., Sivakumar, G. and Xenos, M.A. (2018) 'It's Not Cricket: Examining Political Discussion in Nonpolitical Online Space', *Information, Communication and Society*, 21(11), pp. 1571–87.

Zuckerberg, M. (2009) '200 Million Strong'. Available at https://www.facebook.com/notes/facebook/200-million-strong/72353897130/ (accessed 27 September 2019).

Zuckerberg, M. (2017) 'Building Global Community'. Available at https://www.facebook.com/notes/mark-zuckerberg/building-global-community/10154544292806634/ (accessed 27 September 2019).

16 Citizenship and the data subject

Leah A. Lievrouw

Citizenship is a cornerstone concept in democratic politics and governance. It 'supposes a certain kind of agency, one that involves the individual's capacity to reflect on his/her subjective good as well as on the good of the whole . . . [such] agency has long functioned as a threshold condition for citizenship' (Leydet, 2017). Agency, and thus the capacity for citizenship, are cultivated through communication: in conversation and deliberation with other people, engagement with state institutions and authorities, and participation in the procedures, practices and rituals of democratic rule. Crucially, political agency gives citizens the 'capacity to act independently of authorities and this ability, in turn, depends on whether they regularly act and communicate together, even if this interaction is often mediated through institutions like the electronic media' (Leydet, 2017).

In complex contemporary societies communication media and information technologies are indispensable to political action in all of the senses above, allowing citizens to express their views, interact and access information (though also susceptible to the promulgation of propaganda, suppression of press freedoms, or cultural or political censorship). This role depends on two key assumptions: first, that individuals have some degree of autonomy regarding their self-representations, social relations and interactions (whether in person or via technological communication channels), and their choices of information sources; and second, that media technologies and institutions are (at least in theory) designed to support and extend communicative action and autonomy.

However, what happens when media systems and infrastructure foster expression and communicative action only as a means to capture data about individuals' traits, preferences, beliefs and activities that state and private interests may process and exploit? What happens to public engagement and political action when people's interactions and expressions are reconceived as data streams to be diverted, analysed and predicted for economic, cultural or political gain? Such instrumentality is a common feature of the present media environment, a process recently characterized as 'datafication' (Mayer-Schönberger and Cukier, 2013; van Dijck, 2014). In this chapter I wish to explore this reorientation for people's political communication, autonomy, agency and action; ultimately, for citizenship.

An enormous body of research and scholarship in recent years has examined the uses and implications of digital technologies in politics, particularly since the so-called Great Recession of 2008–09 and the subsequent rise of grassroots political movements and activism around the world using social media platforms (e.g., Castells, 2015; Lievrouw, 2011; Milan, 2013). Digital or social media are widely studied in conventional political parties, campaigning and elections, community deliberation and opinion formation, and governance across community, regional, national and even global contexts (Chadwick, 2006; Howard, 2006; Lutz and du Toit, 2014). Governments' repressive uses of digital technologies to surveil and document their populations in the name of national security, local policing or public welfare have also been investigated (e.g., Eubanks, 2017). The collection of and public access to government information and the rise of open data, open government, civic hacking and data activism movements is another growing research front (Baack, 2015; Gordon and Mihailidis, 2016).

Most of these studies have taken a broadly middle-range (cultures, communities, political movements or campaigns) or macro-level (national or comparative) perspective on collective action, politics and the public. This reflects the scale of online platforms and growing scope of online political expression and movement organization over the last decade. However, the focus here is more on the micro-scale, individual level of analysis, particularly how individual political actors and citizens are conceived, characterized and understand their own agency, within a media context dominated by data capture and predictive analytics. I begin with an overview of how such 'datafied' individuals have been characterized, and then discuss how these characterizations articulate with contemporary notions of democratic citizenship, especially how the citizen's political agency may be shaped, enabled or constrained by media systems based on pervasive data collection and profiling. I consider the European Union's 2018 General Data Protection Regulation (GDPR) as a key policy intervention that invokes its own particular construct, the 'data subject', and which fosters political agency by supporting individuals' privacy. I conclude with a few possible directions or questions for future study.

Data and people

The collection of quantified intelligence about individuals is not new. Ancient empires conducted censuses of subject peoples for taxation; religious authorities systematically recorded births and deaths; aristocratic rulers and bureaucracies compiled legal and property records to document and to assert their legitimacy and authority. Technological systems have radically expanded the scale and scope of information-gathering and record-keeping over time. Early empires established and legitimized their cultural knowledge, and thus extended their influence over time and space, creating 'monopolies of knowledge' (Innis, 1950). In the modern era printing, photography, telegraphy, broadcasting, telecommunications, cinema and video, and digital computing have all been used to collect information about individuals, thereby concentrating and stabilizing the power of those who control the technologies.

By the mid-twentieth century, popular anxieties about widespread surveillance and data collection on individuals had become a common cultural trope, a theme in dystopian fiction (Kafka, 1941; Orwell, 1949). Student protesters in the 1960s condemned data-driven warfare, the militarization and corporatization of academic research, and government surveillance of citizens, particularly political dissidents. The phenomenal growth of networked computing and the Internet that grew out of the military ARPANET (the Advanced Research Projects Agency Network) in the 1980s and 1990s, were seen by some as new opportunities for political voice, inclusion and participation; the decentralized architecture of the Internet was often depicted as inherently democratic (Turner, 2006).

Still other observers saw the same systems as potential threats to personal liberty and well-being. Policy scholar Kenneth Laudon (1986) assessed the systemic social risks of large-scale, computerized personal data collection, which he called the 'dossier society'. He introduced the term 'data image' (Laudon, 1986: 4), defined as 'personal information from a wide variety of public and private sources' that can be combined 'to make life-shaping decisions about individuals with little or no face-to-face interaction, due process, or even knowledge of the individual' in national and global markets for information (Laudon, 1996: 704, n. 7.)

Emphasizing individual subjectivity, digital culture theorist Mark Poster (1990) suggested that a person's 'electronic image' in databases acts an 'additional self, one that may be acted upon to the detriment of the "real" self without the "real" self ever being aware of what is happening' (Poster, 1990: 97–8). For French philosopher Gilles Deleuze (1992), pervasive data collection undermines the very idea of a coherent, singular or independent 'individual', as associated with modernity. Instead, individuals are subdivided, sectioned or parcelled out into countless 'dividuals', according to the purposes of the analyst.

Technology policy and strategy researcher Roger Clarke proposed the concept of 'digital persona' (Clarke, 1994), 'a model of an individual's public personality based on data and maintained by transactions, and intended for use as a proxy for the individual' (ibid.: 78). Such transactional 'signals' have the 'cumulative effect . . . [of] something that approximates personality' (ibid.: 77–8). Digital personae must be attributable to 'a specific, locatable human being' and are 'implicitly assumed . . . to provide a model of the individual that is accurate in all material respects' (ibid.: 84). Creators may exercise power remotely through their personas, but personas may also become 'autonomous' and escape the originator's control, as with a virus that spreads through systems uncontrollably.

Geographer Michael Curry (1997: 692) suggested that data-profiling generates largely unknowable 'digital individuals': 'it is hard to see how any individual today can know whether he or she has adequate knowledge of which data exist, has access to those data, has the ability to correct those data, or can be assured that data have been collected only where necessary'. Profiles are often perceived as being unitary or cohesive impressions of the whole person, but when one's digital individual is invoked, 'I am being treated not like "me" but like a caricature' (Curry, 1997: 694). Nonetheless, they are 'talking for us, representing us, and making decisions for us, [but] we can see that they are very much like the fragmented parts of ourselves that we present in every part of our everyday life' (ibid.: 695).

By the end of the twentieth century *The Economist* noted growing concerns about the rise of global networks and warned of 'the end of privacy' (*The Economist*, 1999), well before the advent of smartphones, social media, big data or the Internet of Things (IoT). Similarly, privacy researcher and advocate Daniel Solove raised serious concerns about 'an electronic collage that covers much of a person's life – a life captured in records, a *digital person* composed in the collective computer networks of the world' (Solove, 2000: 1). Kevin Haggerty and Richard Ericson (2000) argued that proliferating surveillance systems had coalesced into an enormous 'surveillant assemblage', in which disparate data about individuals are 'reassembled into distinct "data doubles" which can be scrutinized and targeted for intervention' (Haggerty and Ericson, 2000: 606). Other scholars elaborated on the data double as 'just one of the endless number of possible combinations [of] fragmented, decontextualized information, collected for many specific purposes, [which] may acquire a multitude of completely different meanings depending on its particular compilation, re-contextualization and application' (Los, 2006: 78; see also Lyon, 2001; Terranova, 2004).

In the wider culture, however, such worries grew quiet with the start of the new century. Spectacular terrorist attacks, and the ruthless national-security response from the targeted states, seemed to justify global-scale surveillance and profiling as part of the 'global war on terror', despite alarms raised by scholars and activists. The interventionist art collective Critical Art Ensemble (CAE) (2001), for example, warned that the collection of electronic records had already reached a stage of 'horrific excess'. For CAE:

> the total collection of records on an individual is his or her data body – a state-and-corporate-controlled Döppelganger . . . the data body not only claims to have ontological privilege, but actually has it. What your data body says about you is more real than what you say about yourself. (Critical Art Ensemble, 2001)

Novel consumer technologies and online services also diverted people's concerns from digital surveillance. The launch of various platforms in the 2000s, tagged with the neologism 'social media' (see Boyd and Ellison, 2007; Wikipedia, https://en.wikipedia.org/wiki/Social_media) encouraged people to display their networks of social relations and contacts publicly, online. The introduction of smartphones that combined mobile telephone service with Internet connectivity (for example, the 'convergent' BlackBerry in 2002 and the first touchscreen iPhone in 2007) promoted 24/7 personal access and connectivity. Devices and platforms normalized 'mass self-communication' (Castells, 2009). Their immediacy and convenience discouraged much careful reflection about how such massive streams of intimate information might be kept, used or passed on, by whom, or why.

Yet, even as people have continued to feed their whereabouts, preferences, choices and routines to a handful of mammoth private firms, privacy worries have returned as trust across public, private, cultural and social institutions has eroded. One surrenders data not only to use credit cards, laptops, phones, toll roads, air travel or delivery services, but also to ubiquitous security cameras, thermostats, light bulbs, vehicles, door bells and grocery store shelves. The proliferation of colloquialisms, such as 'digital breadcrumbs' (Pentland, 2013), 'data cloud', 'digital footprint' and 'data exhaust', suggest the haze or remains of personal information left online as people go about their daily business.

As early as 1994, Roger Clarke described the social types or roles that individuals play within their online communities as data or digital 'shadows'. A decade later, Matthew Zook et al. (2004) noted that the information generated unintentionally or inadvertently in the course of people's routine activities might be especially valuable to commercial and political interests: 'We produce our own data shadow, but do not have full control over what it contains or how it is used to represent us. It has become a valuable, tradeable commodity' (Zook et al., 2004: 169). Phillip Howard applied the concept to political communication online: 'The data shadow follows us almost everywhere. We are not always aware of its appearance, but others can observe our silhouette. The data shadow has become an important political actor . . . What is meaningfully represented in contemporary political institutions is not you but your data shadow' (Howard, 2006, pp. 188–9). The research firm IDC adopted the term 'digital footprint' to denote data generated by users themselves, and 'digital shadow' for data generated by others about users' activities, and found that as of 2008 shadows constituted a larger proportion of all data generated than footprints (Lohr, 2008; Savvas, 2008; see also Kitchin, 2014; Koops, 2011).

The issue of personal data capture reached an important inflection point in the 2010s with the emerging paradigm of big data (Barocas and Nissenbaum, 2014). Dismissed by some as a corporate buzzword, or a difference of degree rather than an entirely new phenomenon, scholars and critics nonetheless acknowledged big data's growing power and problems. In popular discourse, meanwhile, the term suggests the promises (individualized medicine, 'personal assistants' anticipating one's every impulse) and dangers (loss of personal privacy and dignity, robotic warfare) of pervasive data capture. It highlights the persistent, uneasy fit between people's senses of themselves as actually existing persons in the world and the (largely unseen, but highly consequential) profiles and inferences being drawn about them from data: 'a person's data shadow does more than follow them; it precedes them' (Kitchin, 2014: 179).

This process and sensibility have been called 'datafication' (Mayer-Schönberger and Cukier, 2013; see also Breiter and Hepp, 2018; van Dijck, 2014), the digital recording and quantification of virtually every feature and nuance of society and sociality, including qualitative phenomena and subjective experience. Datafication renders life into big data that is collected and deployed for reasons that may be unrelated to any original purpose or context. Indeed, Mayer-Schönberger and Cukier argue that datafication is reorienting knowledge production itself from an epistemology based on inferences from limited, controlled observations to one without uncertainty, based on the discovery of massively complex phenomena directly from 'total' datasets, supposedly without bias or misinterpretation.

Mayer-Schönberger and Cukier have been criticized for their generally non-critical view of datafication. Shoshana Zuboff, in contrast, takes a strong critical position, using the word 'rendition' to describe such wholesale expropriation of people's everyday life into data, to emphasize the word's connotations of surrendering, 'giving up' or 'giving over'. She is clear: *'there can be no surveillance capitalism without rendition'* (Zuboff, 2019: 235).

Finally, one of the more subtle and intriguing recent efforts to theorize the articulation between people and data is Louise Amoore's notion of the 'data derivative' (Amoore, 2011). Amoore uses financial derivatives as the template for this idea, where useful or profitable decisions are based on shifts in value of any component of the assets contributing to the derivative, rather than shifting ownership of the assets in themselves. Most concepts to this point have assumed some degree of mapping between real persons and their representations in data (whether faithful or not). However, a data derivative is not a straightforward profile, but an emergent state generated by the fluid associations across diverse, unknown, partial and constantly reorganizing data elements, which creates a continuously changing horizon or distribution of possibilities. These possibilities are rendered as flags or scores, for example, to make decisions about and interventions in what people do:

> the deployment of a data derivative does not seek to predict the future, as in systems of pattern recognition that track forward from past data . . . because [the data derivative] is precisely indifferent to whether a particular event occurs or not. What matters instead is the capacity to act in the face of uncertainty, to render data *actionable*. (Amoore, 2011: 29)

Here, what is excluded from a given situation is as important as what is classified into it; the objective is not completeness or the reduction of uncertainty, but working with uncertainty to make useful choices. Every new element or decision, or inclusion or exclusion, contributes to a feed-forward of options for what might be, and what decisions might be taken, at any point.

Data and citizens

Clearly, early anxieties about personal data collection and profiling have persisted and become even more acute with the growth of online media. Here, however, we might consider the implications of the various framings of the relationship between people and data for democratic participation and citizenship.

First, most concepts present the relation as a kind of projection, where data are intended to reflect some 'reality' or actual features of the natural person to which they refer. Many writers focus on the accuracy or fidelity of the projection. Most comment on the harms that can arise when incomplete or flawed data are used to construct profiles, that is, the distance or closeness between the data composite and its human referent. Some see flexibility or possibilities for 'degrees of freedom' of action when data are inconsistent or gappy, but others focus on data representations that are too uncomfortably complete, close or faithful to the individual's real characteristics. Indeed, representations that resemble a person's life too closely can have an uncanny quality, uncomfortably tailored to, or intruding on, what is assumed to be private knowledge.

Many concepts also draw sharp distinctions between the presumed coherence and integrity of intact, real individuals, and the idiosyncrasy and variety of data sources that may contribute to a given data representation or profile. Several writers highlight the fundamental unknowability of how so many various and specialized data sources might be recomposed or collaged together to make a representation that the individual might object to, or indeed might not even be aware of.

However, perhaps the most important theme for the present discussion is how data collection and profiling affect people's autonomy and agency. Most accounts view data profiling as a threat to personal autonomy and the individual's freedom from unwarranted surveillance and the inferences drawn from it. It reduces one's ability to form and express independent views, and thus the capacity to act in one's own interests and those of a larger community. The 'invisible' or infrastructural quality of data capture (Star and Bowker, 2006) – most effective when targeted individuals are least aware of it and thus act naturally or authentically – is pivotal here. People's data traces may feed into and shape the very reality and choices for action that they believe are open to them. Put more simply, writers worry that instead of being active agents with control over their self-presentation, persons are being reconceived primarily as data sources, as acted-upon objects subject to data capture.

If data capture threatens individuals' autonomy and agency, and by extension their capacity for political action, we might ask what roles autonomy and agency actually play in contemporary theories of citizenship. In a comprehensive overview of current literature, Leydet (2017) sets out three different (if somewhat interrelated) views: citizenship as legal status; as community membership and source of identity; and as political agency. The first is principally nominal, a designation that confers or withholds certain protections or privileges. The second is associational, based on an individual's affinity with and belonging to a collective. Both of these might require data to make such designations or to determine collective identity. However, the third definition, which emphasizes the individual's capacity to act, may be most relevant here: citizenship entails the ability to interact with others, to reflect on that interaction, and to act (individually or collectively) as a result of this engagement.

Leydet notes that all conceptions 'share the idea that citizenship supposes a kind of agency'. Most political theorists accept implicitly that 'citizens are political agents through their participation in institutions and practices that require significant interaction and mutual awareness'. This includes both 'horizontal' communication with other citizens and peers, and 'vertical' communication with the institutions and representatives of government. Ultimately, 'the capacity for rational agency has long functioned as a threshold condition for citizenship'.

This capacity may be what is most affected by pervasive datafication, which defines and shapes how media systems represent the world to people, including the boundaries and possibilities of action. As noted above, for Leydet the defining quality of citizens as political agents is their 'capacity to act independently of authorities'. Yet if a citizen's agency is induced or constrained by data capture systems, such independence is curtailed too, virtually by definition. One's actions may be limited, for example, by the ubiquitous profiling required for positive identification used in routine economic and social transactions, or by the pervasive simulation and visualization of one's surroundings, resources, options and opportunities. Systems such as Global Positioning System (GPS) mapping, personal fitness monitors or flight trackers may be enormously useful or convenient, but they may also reduce users' abilities to form an independent understanding of the world.

The problem is compounded if individuals are deliberately steered to benefit those who do the monitoring and control the data, rather than their own interests. Continuous data profiling may compel subjects to adapt their behaviour in tandem with system requirements and expectations, leading to a further loss of autonomy and capacity, further profiling and so on. Media anthropologist Veronica Barassi (forthcoming) provides a troubling example. In a recent study of parents and families in the United States, United Kingdom and Europe, she found that children are increasingly subject to data gathering and profiling. She explored the 'narratives' about children that data tell, and the consequences for their future roles as independent adults and citizens: 'the datafication of children . . . not only impact[s] on their right to privacy, but . . . on their right to self-representation, moral autonomy and contextual integrity' (Barassi, forthcoming).

Barassi examines four major contexts of data gathering about children (healthcare, education, home life and social media). Many begin before the child is born, and generate unique individual profiles that persist through the lifespan. Parents are often required or coerced into giving up their children's data, and in interviews expressed a profound sense of helplessness, mixed with resignation or denial, over the normalization of pervasive data collection about their children.

Data profiles cast children primarily as consumers; marketers categorize and discriminate among them and predict their future behaviour and habits, as they do with adults. But Barassi worries that this logic, and the consumer data themselves, are increasingly being appropriated and repurposed by political campaigns and interest groups, leading to what she calls the 'datafied citizen': '*data traces are made to speak for and about* us in public' (Barassi, forthcoming). Children are being conditioned for this future role in 'data environments . . . [that are] exacerbating the lack of agency and autonomy of children in defining themselves as citizen subjects' (ibid).

For Barassi and other scholars, privacy is the *sine qua non* for the cultivation of autonomy and the capacity for self-rule and citizenship. Julie Cohen (2012) argues strongly that, 'Conditions of diminished privacy . . . impair both the capacity and scope for the practice of citizenship' (ibid.: 7), and that such conditions affect the 'material conditions of access to information . . . [which] structure the practice of citizenship' (ibid.: 8). Surveillance is 'a mode of privacy invasion [and] knowledge production designed to produce a particular way of knowing and a mode of governance designed to produce a particular kind of subject . . . [that is] tractable, predictable' (ibid.: 11). Surveillance modulates action, where modulation is 'a set of practices in which the quality and content of surveillant attention is [sic] continually modified according to the subject's own behavior . . . [and] logics that ultimately are outside the subject's control' (ibid.: 10). Cohen warns that the 'habits that support a mature, critical subjectivity . . . require privacy to form. The institutions of modulated democracy, which systematically eradicate breathing space for dynamic privacy, deny both critical subjectivity and critical citizenship the opportunity to flourish' (ibid.: 12).

Other scholars are more sceptical about privacy as the main guarantor of autonomy. Solon Barocas and Helen Nissenbaum (2014) argue that privacy has become something of a hollowed-out term: the public and policy-makers alike now tend to attribute virtually any discomfort, risk or exposure of people's online activities to the violation of privacy rights. Beyond privacy, they say, anonymity is necessary for intellectual freedom and political agency today;

but it is being eradicated by platforms' restrictive terms of service, big data compilation and machine learning, as even digital voting systems, with their well-known vulnerabilities to hacking, are adopted by ever more jurisdictions.

Nonetheless, privacy has become the cornerstone of new regulatory regimes that attempt to ameliorate the exploitation of people and their data. One of the most notable is the European Union's 2018 General Data Protection Regulation (GDPR). Its core principle is the protection of the 'data subject', the:

> identified or identifiable natural person . . . who can be identified, directly or indirectly, in particular by reference to an identifier such as a name, an identification number, location data, an online identifier or to one or more factors specific to the physical, physiological, genetic, mental, economic, cultural or social identity of that natural person (GDPR, Article 4 (1))

Specifically, the GDPR enumerates the rights of data subjects and principles of data protection that pertain to those subjects and data about them. It defines and sets the limits or conditions for personal data processing and profiling by entities that determine what and how such processing is done (controllers). It is especially important for this discussion because it arose out of rising concerns about the unprecedented power of state and non-state organizations using global digital networks to gather data and to profile and influence individuals' behaviour, wherever they are and whatever they happen to be doing.

Prior to the GDPR, such data was generally obtained without the subject's explicit consent or clear understanding of how tracking works, or how the data is ultimately used, by whom, for what purposes – that is, the very situation of disjuncture and risk to personal autonomy that has generated so many concepts such as data shadow, data image, digital persona, digital individual, data double, and so on. Of course, these notions attempt to describe the nature of a person's representation in data, and so differ from the GDPR's data subject, the 'identifiable natural person' being represented. But this distinction is also important with respect to citizenship because it reorients the analytic and policy focus back towards people as actors, rather than the constant turnover of miscellaneous data assembled from scattered sources.

Citizens and data subjects are similar in some respects. Both are understood first as natural persons with autonomy, agency and a capacity for understanding and action on behalf of themselves and their communities. Both may have the capacity to act independently of state interests. However, citizens differ from data subjects in terms of their relation to the state. Citizenship is just one, if a special one, of the 'factors specific to the . . . cultural or social identity of that natural person', as stated in Article 4 of the GDPR: a legal status or designation. Citizens' political activities are customarily protected by privacy or anonymity, precisely to ensure their autonomy and independence. Data subjects, in contrast, are defined primarily on the basis of being 'identifiable' as a consequence of extensive data collection and processing; they are resources whose actions and characteristics are recorded and 'datafied', rather than political actors whose autonomy and agency must be protected. Put differently, not all data subjects are citizens. But citizens are increasingly likely to be data subjects.

Debates about the GDPR and its implementation have been framed largely in terms of privacy protection for individuals, in response to popular misgivings about pervasive personal data collection, and the motives of the institutions and enterprises that conduct it. Likewise, its requirements and remedies presume the active consent of the individual data subjects. Certainly, privacy is desirable in itself as a condition of all people's well-being and dignity. However, it can be argued that the ultimate value of privacy for democratic societies (and why the European Union's member states would seek to protect it), is in the formation of the capable political subject or citizen. As scholars cited here have pointed out, privacy is a necessary condition for the cultivation of people's political agency, their ability to reflect on their own and the common good, and especially their capacity for communication and deliberation with others about issues that affect them; what political philosopher Jürgen Habermas has termed collective 'will-formation' (Habermas, 1996).

Ways ahead: citizens, data subjects

My aim in this chapter has been to point out, first, the long and uneasy tension between the ways that people conceive of themselves as natural persons who exist and act in the world, and the proliferation of digital information and media systems that capture and process personal information on a massive scale, and render data composites or profiles that are presumed to stand as effective proxies for those individuals. I have shown how observers from various disciplines and perspectives have described that tension and its consequences, mainly in terms of alternate, parallel or shadow selves. Principal among these consequences are those for individual autonomy, agency and citizenship, as the conditions for cultivating personal capacity, enquiry, deliberation, opinions and collective political 'will-formation' are gradually surveilled and coded away. I have suggested that the European Union's GDPR, one of the most significant recent efforts to regulate the widening landscape of relations between people and the data they generate, is important not just as a privacy regime, but also because such privacy protections also foster individuals' autonomy, agency and capacity for citizenship. In line with this volume's theme, then, I would like to suggest some possible questions or avenues for further investigation into the prospects for citizenship, as digital mediation becomes the defining condition of everyday life and experience.

The first and most basic question regards the nature of citizenship itself in technologically advanced societies. Is it merely a legal designation, social affiliation or list of traits? Are citizens, like other data subjects, anything more than aggregations of data points ready to be engineered (for example, the infamous Cambridge Analytica case in the 2016 United States presidential election)? By 'digital citizenship', some writers mean the straightforward migration of traditional political practices (voting, campaigning, mobilizing, issue advocacy) to digital tools and platforms, particularly as taught in elementary and secondary civics curricula (e.g., Rogers-Whitehead, 2019). Other scholars, such as Julie Cohen, have made important strides toward understanding the threats to democratic participation associated with surveillant media. But we might ask whether the nature of participation or the rights, responsibilities and civic culture that cultivate citizenship have changed in any fundamental ways as a result of pervasive digital mediation.

This raises another fundamental question: what conditions are necessary for individuals to cultivate their capacity and agency as citizens? Those conditions have long been associated with spaces and opportunities for social engagement and deliberation, encountering new or different viewpoints, and developing independent thought and civility; what Helen Nissenbaum (1998) has called 'being private in public'. However, the context of engagement in person is substantially different from that online. (Even the earliest studies of computer-mediated communication showed that the impersonal quality and 'reduced social context cues' of online interaction can have 'substantial deregulating effects' on people's communicative behavior (Sproull and Kiesler, 1986: 1492). If nature abhors a vacuum, as the saying goes, today's Internet abhors personal privacy and especially anonymity; ostensibly, the necessary conditions for independent thought and liberty. What, if anything, might take their place?

A third line of questions asks how citizens-as-data-subjects fit into what many contemporary scholars think of as 'governmentality', the nature of how citizens may be governed or secure and participate in their own self-governance (Huff, 2007). Leydet (2017) outlines two conventional models of citizenship with respect to governance: a 'republican' model based on strong, direct self-rule, where each citizen has the capability and indeed the obligation to participate and rule alongside their peers; and a 'liberal' model based on citizenship as a legal status, which emphasizes the individual's personal liberty, including freedom from the obligations of rule, and thus the delegation of political duties and power to representatives. In complex, developed societies the latter is by far the norm, but for citizens-as-data-subjects, is this model of political or civic citizenship even sustainable? Are there different possibilities for civic engagement or political formations – themselves mediated, perhaps – that would increase citizens' chances to participate and share power, not just surrender it on demand along with their data traces?

In sum, in recent years many observers have been quick to abandon the early, idealistic visions of the Internet as an emancipatory, inclusive forum for democratic participation. They ascribe the current climate of political uncertainty, reaction and erosion of trust in democratic, pluralist institutions across the world to the relentless, ubiquitous growth of digital platforms, increasingly left to the control of private sector monopolists and repressive authoritarian states. This bleak prospect may yet be refuted, but that will require a sharp reassessment of the principles and practices of citizenship that begins, but cannot end, with the data subject.

References

Amoore, L. (2011) 'Data Derivatives: On the Emergence of a Security Risk Calculus for Our Times', *Theory, Culture and Society*, 28(6), pp. 24–43.

Baack, S. (2015) 'Datafication and Empowerment: How the Open Data Movement Re-Articulates Notions of Democracy, Participation, and Journalism', *Big Data and Society*, July–December, pp. 1–11. Available at https://doi.org/10.1177/2053951715594634/ (accessed 13 November 2019).

Barassi, V. (forthcoming) *Child | Data | Citizen*. Cambridge, MA: MIT Press.

Barocas, S. and Nissenbaum, H. (2014) 'Big Data's End Run Around Anonymity and Consent'. In J. Lane, V. Stodden, S. Bender and H. Nissenbaum (eds), *Privacy, Big Data, and the Public Good: Frameworks for Engagement*. New York: Cambridge University Press, pp. 44–75.

Boyd, D. and Ellison, N.B. (2007) 'Social Network Sites: Definition, History, and Scholarship', *Journal of Computer-Mediated Communication*, 13(1), pp. 210–30.

Breiter, A. and Hepp, A. (2018) 'The Complexity of Datafication: Putting Digital Traces in Context'. In A. Hepp, A. Breiter and U. Hasebrink (eds), *Communicative Figurations: Transforming Communications in Times of Deep Mediatization*. Cham: Palgrave Macmillan, pp. 387–406.

Castells, M. (2009) *Communication Power*, 1st edn. Oxford, UK and New York, USA: Oxford University Press.

Castells, M. (2015) *Networks of Outrage and Hope: Social Movements in the Internet Age*, 2nd edn. Cambridge: Polity.

Chadwick, A. (2006) *Internet Politics: States, Citizens, and New Communication Technologies*. New York, USA and Oxford, UK: Oxford University Press.

Clarke, R. (1994) 'The Digital Persona and Its Application to Data Surveillance', *Information Society*, 10(2), pp. 77–92.

Cohen, J.E. (2012) 'What Privacy Is For', *Harvard Law Review*, 126, 1904–33.

Critical Art Ensemble (CAE) (2001) 'The Mythology of Terrorism on the Net'. In *Digital Resistance: Explorations in Tactical Media*. New York: Autonomedia, pp. 31–40.

Curry, M.R. (1997) 'The Digital Individual and the Private Realm', *Annals of the Association of American Geographers*, 87, pp. 681–99.

Deleuze, G. (1992) 'Postscript on the Societies of Control', *October*, 59, pp. 3–7.

The Economist (1999) 'The End of Privacy', 1 May, p. 15.

Eubanks, V. (2017) *Automating Inequality: How High-Tech Tools Profile, Police, and Punish the Poor*. New York: St Martin's Press.

Gordon, E. and Mihailidis, P. (eds) (2016) *Civic Media: Technology | Design | Practice*. Cambridge, MA and London: MIT Press.

Habermas, J. (1996) *Between Facts and Norms: Contributions to a Discourse Theory of Law and Democracy* (trans. W. Rehg). Cambridge, MA: MIT Press.

Haggerty, K.D. and Ericson, R.V. (2000) 'The Surveillant Assemblage', *British Journal of Sociology*, 51(4), pp. 605–22.

Howard, P.N. (2006) *New Media Campaigns and the Managed Citizen*. Cambridge, UK and New York, USA: Cambridge University Press.

Huff, R.F. (2007) 'Governmentality'. In M. Bevir (ed.), *Encyclopedia of Governance*. Thousand Oaks, CA: SAGE, pp. 389–90.

Innis, H.A. (1950). *Empire and Communications*. Oxford: Clarendon Press.

Kafka, F. (1941). *The Castle*, 2nd edn (trans. E. and W. Muir, with an introduction by Thomas Mann). New York: Alfred A. Knopf.

Kitchin, R. (2014). *The Data Revolution: Big Data, Open Data, Data Infrastructures and their Consequences*. London, UK and Thousand Oaks, CA, USA: SAGE.

Koops, B.-J. (2011) 'Forgetting Footprints, Shunning Shadows: A Critical Analysis of the "Right To Be Forgotten" in Big Data Practice', *Script/ed*, 8(3), pp. 229–56.

Laudon, K. (1986) *The Dossier Society: Value Choices in the Design of National Information Systems*. New York: Columbia University Press.

Laudon, K. (1996) 'Markets and Privacy'. In R. Kling (ed.), *Computerization and Controversy*. San Diego, CA, USA and London, UK: Academic Press, pp. 697–726.

Leydet, D. (2017) 'Citizenship'. In E.N. Zalta (ed.), *Stanford Encyclopedia of Philosophy*. Available at https://plato.stanford.edu/archives/fall2017/entries/citizenship/ (accessed 13 November 2019).

Lievrouw, L.A. (2011) *Alternative and Activist New Media*. Cambridge: Polity.

Lohr, S. (2008) 'Measuring the Size of Your Digital Shadow'. *New York Times*, 11 March. Available at https://bits.blogs.nytimes.com/2008/03/11/measuring-the-size-of-your-digital-shadow.

Los, M. (2006) 'Looking Into the Future: Surveillance, Globalization and the Totalitarian Potential'. In D. Lyon (ed.), *Theorizing Surveillance: The Panopticon and Beyond*. London, UK and New York, USA: Routledge, pp. 69–95.

Lutz, B. and du Toit, P. (2014) *Defining Democracy in a Digital Age*. Basingstoke: Palgrave Macmillan.

Lyon, D. (2001) *Surveillance Society*. Buckingham, UK and Philadelphia, PA, USA: Open University Press.

Mayer-Schönberger, V. and Cukier, K. (2013) *Big Data: A Revolution that will Transform How We Live, Work, and Think*. New York: Houghton Mifflin Harcourt.

Milan, S. (2013) *Social Movements and Their Technologies: Wiring Social Change*. Basingstoke: Palgrave Macmillan.

Nissenbaum, H. (1998) 'Protecting Privacy in an Information Age: The Problem of Privacy in Public', *Law and Philosophy*, 17, pp. 559–96.

Orwell, G. (1949) *Nineteen Eighty-Four: A Novel*, 1st American edition. New York: Harcourt Brace.

Pentland, A. (2013) 'The Data-Driven Society', *Scientific American*, 309(October), pp. 78–83.

Poster, M. (1990) *The Mode of Information: Poststructuralism and Social Context*. Chicago, IL: University of Chicago Press.

Rogers-Whitehead, C. (2019) *Digital Citizenship: Teaching Strategies and Practices from the Field*. Lanham, MD, USA and London, UK: Rowman & Littlefield.

Savvas, A. (2008) 'Digital Shadow Outstrips Digital Footprint', *ComputerWeekly*, 12 March. Available at https://www.computerweekly.com/news/2240085343/Digital-shadow-outstrips-digital-footprint.

Solove, D.J. (2000) *The Digital Person: Technology and Privacy in the Information Age*. New York, USA and London, UK: New York University Press.

Sproull, L. and Kiesler, S. (1986) 'Reducing Social Context Cues: Electronic Mail in Organizational Communication', *Management Science*, 32(11), pp. 1492–1512.

Star, S.L. and Bowker, G. (2006) 'How to Infrastructure'. In L.A. Lievrouw and S. Livingstone (eds), *Handbook of New Media*, updated student edn. London: SAGE, pp. 230–45.

Terranova, T. (2004) *Network Culture: Politics for the Information Age*. London, UK and New York, USA: Pluto Press.

Turner, F. (2006) *From Counterculture to Cyberculture: Stewart Brand, the Whole Earth Network, and the Rise of Digital Utopianism*. Chicago, IL: University of Chicago Press.

Van Dijck, J. (2014) 'Datafication, Dataism and Dataveillance: Big Data between Scientific Paradigm and Ideology', *Surveillance and Society*, 12(2), pp. 197–208.

Zook, M., Dodge, M., Aoyama, Y. and Townsend, A. (2004) 'New Digital Geographies: Information, Communication, and Place'. In S.D. Brunn, S.L. Cutter and J.W. Harrington Jr (eds), *Geography and Technology*. Dordrecht: Kluwer Academic Publishers, pp. 155–76.

Zuboff, S. (2019). *The Age of Surveillance Capitalism: The Fight for a Human Future at the New Frontier of Power*. New York: Public Affairs.

17 Citizens and their political institutions in a digital context

Elizabeth Dubois and Florian Martin-Bariteau

Introduction

Political systems are unavoidably digital. From digital government service delivery, to online mobilization for protests in the streets, to disinformation spread across social media and messaging applications, the digital aspects of civic life are inescapable. Even for those few who opt out of using digital tools or those who do not have the access or skills needed, digital politics is so pervasive that they cannot avoid it. With a turn towards smart cities, digital surveillance, and the rise of machine learning and artificial intelligence, everyone and every institution is impacted.

From this perspective, all citizenship has become digital. We aim to consider the ways in which individuals enact their citizenship, from requesting government services to signing petitions, and much more in between. We also consider the ways in which other actors treat individuals as citizens, such as journalists reporting public opinion, data being collected to prioritize government resource deployment, or political parties targeting messages to potential supporters. In these and many other ways, citizenship is digital.

We begin by defining digital citizenship. Next, we offer a research agenda for understanding this citizenship in its digital context by examining relationships between citizens and three political institutions: legal systems, government and political information systems. These relationships existed before digital technology became pervasive, but have evolved as we integrate technology into our daily lives. We focus on research which is most needed to understand the individual's role in the wider political system. Our aim is not to tease out what is an expansive array of future research options in each relationship, but instead to point specifically to research opportunities that can support the development of better policies, tools and frameworks. Notably we exclude partisan politics, private sector and civil society, all of which are crucial components of the political system which other chapters in this book address.

Understanding citizenship in a digital context

Definitions of citizenship vary, but core commonalities can guide our understanding of who is a citizen and what enacting citizenship looks like. Citizenship from a legal perspective is bound by recognition from a given nation-state: 'the notion of citizenship provides people living in these nation-states with certain civil, social, political, and economic rights and responsibilities' (Choi, 2016: 3). Marshall (1964) describes civil (such as freedoms and rights), political (such as ability to exercise power over elected officials, voting) and social (such as civic culture and national heritage) elements of citizenship. These are widely accepted as a base framework for understanding citizenship (Banks, 2008). However, legal conceptualizations can be exclusionary based on factors such as language, religion, ethnic or cultural groups. In contrast, citizenship might be usefully considered in terms of identity and belonging to a community (Banks, 2008; for a discussion of various conceptualizations, see Choi, 2016: 4–5). We adopt a broad definition of citizenship in contrast to narrower legal ones. We use the term 'citizenship' to describe the civic experiences of individuals. Notably, our work focuses on those living within Western democratic states.

Scholarship has often distinguished experiences of online and offline citizenships, conceptualizing digital citizenship as distinct from traditional offline citizenship. Some have conceptualized digital citizenship in terms of behavioural norms online (Ribble, 2004), or in terms of frequency and competence with technology use (Mossberger et al., 2008). Others focus on identifying digital versions of analogue acts of citizenship in order to expand our understanding of what constitutes such an act (Bennett et al., 2011). Similarly, digital citizenship has been construed as any online participation (Greffet and Wojcik, 2014) – as understood by the Council of Europe (Council of Europe, 2018). Hintz and colleagues suggest commonalities across definitions of digital citizenship wherein, 'in most iterations it tries to understand the relation between the digital and the political, and thus the role of the digital subject as a political subject' (Hintz et al., 2019: 20). They argue, commenting on the role of government and business in capitalist systems, 'Digital citizenship thus denotes our roles, positions and activities in a society that is organized through digital technologies' (ibid.: 144).

Problematically, many definitions of digital citizenship require active and intentional use of digital technology by individuals. Research into digital divides has shown that individuals around the world may choose not to use digital tools, may not have access to the Internet and/or may not have the skills required to make use of tools. This does not mean that their experience of citizenship is not digital. Regardless of the individual citizens' digital activity, policies and laws about digital data, privacy and other issues are developed; data about individuals is collected and used by governments; and political information is almost always shared digitally at some point in the system. The result is a rich array of policy and practical challenges for political players, as well as opportunities for future research. The following sections examine specific relationships between citizens and three political institutions: legal systems, government services and political information systems.

Supporting fundamental freedoms in a digital context

Governed by algorithms which are often obscure (Pasquale, 2015), the digital society raises multiple legal, political and ethical issues. Constant connectivity, the platform economy, smart cities and other digitalizations of everyday civic acts bring into question the effectiveness of the law in the face of autonomous systems (Hildebrandt, 2016). This calls for the development of frameworks to ensure respect for fundamental rights in the digital space. The ubiquity of algorithms offers opportunities for research into the societal impacts of technologies and how legal systems can effectively and coherently protect citizens in this digital context.

First, given the importance of data in an algorithmic society, research to understand expectations around consent and to shape a coherent framework to protect privacy is needed. Researchers should study methods for enforcing informed, transparent and accessible consent practices as well as better privacy-empowering legal frameworks. To avoid state and corporate surveillance – whether actual or perceived – it is necessary for individuals to have a solid foundation of principles of data security and privacy, and knowledge of how to navigate platform features. In order to address these emerging concerns, public information is needed to increase digital literacy regarding citizens' understanding of privacy risks, the life cycle of data, quality of consent, alternative choices and the segregation of data to ensure privacy.

Second, the rise of the platform economy (Kenney and Zysman, 2016) calls for discussion about the roles, responsibilities and obligations of corporations which provide services such as social networking, search and video on demand. To build better legal responses, we need to understand the changed balance of powers between citizens, political institutions and platforms. The use of private platforms by governments to bring digital services and information to citizens also highlights important research questions on the extension of the government sphere, and how constitutional concerns could arise. Similarly, the delegation of certain government functions to corporate actors, notably for content moderation, moderation of harmful speech, and right of erasure, surfaces concerns related to freedom of expression and information.

Third, it is important to know more about tools used by governments and corporations. Algorithmic societies must establish clear guidelines for protecting civil rights. To create policies which protect citizen rights, policy-makers and citizens need to be able to access information about how digital tools work to enable these changes, from the ones used in the justice system to the ones used to make social benefits assessment. Moreover, research is needed to inform the conceptualization of ethical design frameworks.

Fourth, research to better understand whistleblowing and ethical hacking is needed so that appropriate legal and political frameworks can be designed. The protection of whistleblowers is essential to ensure an open, fair and safe society, and to safeguard citizens' digital rights and security (Martin-Bariteau and Newman, 2018). Whistleblowing protection also has impacts on news media in terms of the protection of journalistic sources, which in turn has an impact on freedom of information and political information systems. Similarly, research is needed around forms of 'hactivism' in pursuit of a socially accepted cause, which often evolve in legally and morally ambiguous spaces (Maurushat, 2019). It is important to strike a fair balance between the public interest and the necessary protection of some governmental and corporate secrets.

Fifth, with the ubiquity of the global and digital economy, research into regulation of platforms and algorithms considering cross-jurisdictional and cross-cultural issues is needed. While most regulations are made at national level, reflecting local values, there are bound to be conflicts over what different states want to regulate. Those states may not share the same core societal values. This calls for comparative studies to design adequate tools and frameworks to tackle this.

An interdisciplinary approach, from law to the social and computer sciences, is needed to better tackle these challenges, to design technology-independent regulations, to develop human rights-centred digital tools, and to work with government and private organizations to effect change at the conception or platform level. Protection of citizens' digital rights is of the utmost importance, and necessary to build trust in a digital society. This will only be possible through coherent and inclusive evidence-based tools and frameworks. From the use of proprietary and unsafe systems, to the delegation of power to private platforms, the protection of citizens' digital rights relies importantly on how government and political institutions are built and designed in a digital context, from data policies to tool design.

Designing government services in a digital context

Citizens interact with their governments intentionally, for example when they seek out services; and inadvertently, for example when governments collect data about Internet penetration. Citizens' practices, preferences and needs must inform the extent to which services could – or should – be digitized, as well as the types of data that governments should gather about citizens. In order to design government approaches this way, more information is needed about citizens' needs, and about how governments currently succeed and fail as they incorporate digital technology.

First, resources need to be created to help governments choose between digital and analogue service delivery, and research is required to develop, test and evaluate them. Analogue services are sometimes better than digital ones, and the perceived promises of digital services – often reduced cost – are not necessarily true or of most value. A more nuanced view of when digitization is worthwhile, keeping in mind the limited digital connectivity in different regions, is needed. There needs to be further study of the economic, social and political costs and benefits of government digitization.

Second, how silos between institutions hinder digital government service delivery merits study. Governments in most Western democracies exist at multiple levels and are divided across different departments. Typically, the type of work, as well as the organizational culture, has impacts on the success of digitization. To develop efficient and effective processes and to understand how digital era government is (or is not) playing out, it is crucial to study specific cases and understand where there are networks of support, versus silos which can be stagnant.

Third, research into who is currently excluded and where current approaches succeed and fail is needed. This can help to create design practices which are both inclusive and centred around accessibility. Too often, private corporations and government alike design tools which work for the majority of a population, but systematically exclude portions of the population who are differently abled, lack certain resources or have no access to opportunities to gain relevant skills. Governments should not exclude groups, and have a responsibility to prioritize accessibility to their services. Research is needed to understand what is, or is not, technically possible, practically useful and ethically desirable. Research into who is currently excluded and why, in specific cases of existing government services, can help to inform future design and development of assessment tools. This will help to track progress and support development of a more nuanced understanding of what technology use looks like across a diverse array of individuals.

Fourth, the feasibility and obstacles to the implementation of an 'open by design' framework for digital government services and data need to be studied. Open data is at the core of open government culture (Open Government Partnership, 2011) promoting collaboration and trust with political institutions. Allowing greater transparency and accountability, public data should be freely available to everyone in a systematic, unrestricted and interoperable way. Open by design promotes a paradigm change in service design, asking, at conception, how data will be released. Beyond data, as the use of proprietary solutions and administrative black boxes raises concerns from the public, research should study how to extend an open by design approach to digital services and tools themselves. In particular, increased use of algorithmic decision-making for the administrative and judicial functions calls for such transparency (see, e.g., Government of Canada, 2019).

Fifth, research is needed into the effectiveness of citizen–government collaborations, into uses of government open datasets, and into civic technology initiatives such as those which embed technologists into government departments. Empowering the public to help create and design digital services and policies can lead to more inclusive and responsive outcomes. The approach can also create a prime environment for natural experiments and case studies, as well as deeply qualitative work which can help us to better understand the changing relationships between citizens and their governments in a digital context.

There is substantial opportunity for cross-national comparison and support. A commonality across these points is experimentation and contextual knowledge: to develop suitable government services and to develop deeper understandings of citizens' relationships with their governments, it is increasingly important to understand what works and what does not in context. This requires a high level of transparency where government will need to let academic researchers in to observe and inform practices. While this may be perceived as risky, it is crucial to the further development of digital services and policies in governments. Similarly, the political context in which governments implement digital changes cannot be ignored. Political will and resources are required for these types of initiatives.

Understanding political information in a digital context

Journalism plays an important role in democratic systems by informing the general public about their political system and current affairs, as well as by holding government to account and by shedding light on parts of the political system that the individual citizen may otherwise not have access to. In a digital media environment, characterized by hybridity, news media are joined by other actors who can produce and disseminate political information, sharing tools and tactics (Chadwick, 2013). The citizen is then powerful in different ways, and can interact with and create political information in new ways such as innovative forms of digital journalism and citizen journalism, as well as media manipulation and disinformation campaigns (Hermida, 2016). Research is needed to build better creation and delivery approaches for political information, be it supporting established forms of journalism or promoting new approaches to informing the public.

First, it is important to know more about how individuals consume political information. While news is a key way in which citizens can learn about political issues, it is far from the only source (Dutton et al., 2017), and therefore we use the term 'political information'. When asking questions about how political information spreads and is consumed we must consider the changing array of types of content. This includes traditional news as well as political information embedded in other types of content, such as political memes, which spread across social media, instant messaging and other platforms.

Second, it is important to establish a distinction between news literacy, media literacy and digital literacy. There has been substantial interest in how to assess the quality of content, how to identify trusted and trustworthy sources of political information, and how to respond to disinformation. News, media and digital literacy are all often pointed to as a key response. But these are different things, which are not consistently defined. We need to distinguish between them and measure them separately.

Third, we need to understand political information processes in a cross-national context. The annual Reuters Digital News Reports have provided one of the most encompassing datasets for comparison; these had covered 38 countries by 2019 (Reuters, 2019). More needs to be done to use these datasets and to develop other large-scale datasets which offer the opportunity to track trends across context, develop and question the applicability of theories in different national and/or cultural contexts, and consider other variables which may be relevant to how information is consumed.

Fourth, an examination of innovative journalistic efforts is needed. A database of case studies and examples of experimental approaches to creating and sharing political information could be helpful for understanding citizens' habits and preferences as well as for developing sustainable business models. Similarly, finding ways to promote solution-oriented journalism and to track progress is crucial. This requires development of metrics of success which can be shared publicly in order to replicate assessment. Solutions-oriented approaches can help us to design new approaches, understand the needs of people and offer citizens information they can act on.

Fifth, understanding which voices are invited, amplified and heard is crucial. We know there are problems related to representation and exclusion in news media traditionally. We also know that local news media and niche outlets struggle to find a sustainable business model in the Silicon Valley 'will it scale?' mentality. We need to understand which voices are present and then find ways to promote diversity.

Across each of these areas for future research that we have identified, an investigation into specifically who is excluded and who is included merits meaningful investigation. This is because political information environments can be siloed and, while people may make efforts to seek out alternative viewpoints (Dubois and Blank, 2018), those with the most resources and existing political interest are advantaged while others may lack capacity.

Conclusion

Across each of the citizen–institution relationships we examined there are commonalities: the need for contextual understanding and, where possible, experimentation; the need for cross-national research; and the need for research which explicitly and intentionally examines exclusion and inclusion of individuals who are not the 'average' users of given digital tools. To do this we need increased efforts to make existing datasets available, construction of datasets which can more easily be joined or compared, and coordination by researchers when collecting and analysing these data. While this might apply to any social science research topic, we think it is particularly relevant for understanding citizenship in a digital context.

Citizenship can be experienced differently depending on a variety of factors. This includes the ways in which individuals relate to their political system, their civic acts, and the way their behaviours and needs might inform the approaches of various political actors. Survey data, behavioural trace data and qualitative data can each help to inform our understanding of how citizenship is experienced in different ways. The triangulation of data types can help us to tease out relevant variables in order to test old theories, develop new theories, and support the creation and implementation of uses of digital technology by political institutions. Furthermore, since the experience of citizenship we have described is not necessarily linked to belonging to a nation-state, cross-national, cross-cultural and other comparative approaches across contexts will be fruitful for understanding when geographic bounds matter. In order to conduct these studies, significant attention to creating datasets which can be compared is needed.

Publishing open datasets is a good start, but detailed labelling of datasets and inclusion of information about how and why they were created in that way is essential. While ethnographers and qualitative researchers are experienced in describing the way in which their data was collected, organized and treated in great detail, quantitative approaches tend to offer less nuance in published studies. Initiatives surrounding the ethics and governance of artificial intelligence (AI) point to the need for detailed tracking of how datasets are constructed and manipulated (Alade et al., 2019; Holland et al., 2018). These kinds of initiatives may best be supported by a consortium approach wherein multiple groups worldwide contribute their knowledge, experience and datasets. Think tanks or university-based research institutes can also support this kind of work and could be used to connect communities, promote opportunities and facilitate the sharing of datasets.

Ultimately, the relationships between citizens and their legal system, their government and their political information systems are necessarily impacted upon by digital technologies. Data, ideally comparative, about citizens' uses and preferences related to digital technologies are needed to understand how individuals enact their citizenship, and how we can develop better policies, tools and evaluation frameworks around those acts. The agenda we put forward is ambitious. We expect that political and economic factors will limit the extent to which it can be implemented. Yet we find it important to lay out ambitious goals in order to highlight possibilities and to promote progress.

Acknowledgements

This chapter builds from a previously published report (Dubois and Martin-Bariteau, 2018) and has been made possible thanks to the support of the Social Sciences and Humanities Research Council of Canada (Connection Grant 611-2016-0538, and Insight Development Grant #430-2018-0972). The authors thank Marie-Hélène Cassimiro and Catherine Ouellet for their research assistance.

References

Alade, Y., Kaeser-Chen, C., Dubois, E., Parmar, C. and Schüür, F. (2019) *Towards Better Classification*. Available at https://kaleidoscope.media.mit.edu/white-paper/ (accessed 20 August 2019).

Banks, J.A. (2008) 'Diversity, Group Identity, and Citizenship Education in a Global Age', *Educational Researcher*, 37, pp. 129–39.

Bennett, W.L., Wells, C. and Freelon, D. (2011) 'Communicating Civic Engagement: Contrasting Models of Citizenship in the Youth Web Sphere', *Journal of Communication*, 61(5), pp. 835–56.

Chadwick, A. (2013) *The Hybrid Media System: Politics and Power*. Oxford: Oxford University Press.

Choi, M. (2016) 'A Concept Analysis of Digital Citizenship for Democratic Citizenship Education in the Internet Age', *Theory and Research in Social Education*, 44(4), pp. 565–607.

Council of Europe (2018) 'Digital Citizenship and Digital Citizenship Education'. Available at https://www.coe.int/en/web/digital-citizenship-education/digital -citizenship-and-digital-citizenship-education/ (accessed 20 August 2019).

Dubois, E. and Blank, G. (2018) 'The Echo Chamber is Overstated: The Moderating Effect of Political Interest and Diverse Media', *Information, Communication and Society*, 21(5), pp. 729–45.

Dubois, E. and Martin-Bariteau, F. (2018) *Canadians in a Digital Context: A Research Agenda for a Connected Canada*. Ottawa, ON: University of Ottawa. Available at https://ssrn.com/abstract=3301352/ (accessed 20 August 2019).

Dutton, W.H., Reisdorf, B., Dubois, E. and Blank, G. (2017) 'Search and Politics: The Uses and Impacts of Search in Britain, France, Germany, Italy, Poland, Spain, and the United States'. Available at https://ssrn.com/abstract=2960697/ (accessed 20 August 2019).

Government of Canada (2019) 'Directive on Automated Decision-Making'. Available at https://www.tbs-sct.gc.ca/pol/doc-eng.aspx?id=32592/ (accessed 20 August 2019).

Greffet, F. and Wojcik, S. (2014) 'La citoyenneté numérique: Perspectives de recherche', *Réseaux*, 184–5(2), p. 125.

Hermida, A. (2016) *The New Information Power-Brokers: Gatekeeping in Hybrid Digital Media*. Vancouver, BC: UBC. Available at http://www.alfredhermida.me/wp-content/uploads/2019/07/2016-sshrc-ksg-hermida_0.pdf/ (accessed 20 August 2019).

Hildebrandt, M. (2016) *Smart Technologies and the End(s) of Law: Novel Entanglements of Law and Technology*. Cheltenham, UK and Northampton, MA, USA: Edward Elgar Publishing.

Hintz, A., Dencik, L. and Wahl-Jorgensen, K. (2019) *Digital Citizenship in a Datafied Society*. Cambridge: Polity Press.

Holland, S., Hosny, A., Newman, S., Joseph, J. and Chmielinski, K. (2018) *The Dataset Nutrition Label: A Framework to Drive Higher Data Quality Standards*. arXiv:1805.03677/ (accessed 20 August 2019).

Kenney, M. and Zysman, J. (2016) 'The Rise of the Platform Economy', *Issues in Science and Technology*, 32(3). Available at https://issues.org/the-rise-of-the-platform-economy/ (accessed 20 August 2019).

Marshall, T.H. (1964) *Class, Citizenship, and Social Development: Essays of T.H. Marshall*. Westport, CT: Greenwood.

Martin-Bariteau, F. and Newman, V. (2018) *Whistleblowing in Canada. A Knowledge Synthesis Report*. Ottawa, ON: University of Ottawa. Available at https://ssrn.com/abstract=3111851/ (accessed 20 August 2019).

Maurushat, A. (2019) *Ethical Hacking*. Ottawa, ON: University of Ottawa Press.

Mossberger, K., Tolbert, C.J. and McNeal, R.S. (2008) *Digital Citizenship: The Internet, Society, and Participation*. Cambridge, MA: MIT Press.

Open Government Partnership (2011) 'Open Government Declaration'. Available at https://www.opengovpartnership.org/process/joining-ogp/open-government-declaration/ (accessed 20 August 2019).

Pasquale, F. (2015) *The Black Box Society: The Secret Algorithms That Control Money and Information*. Cambridge, MA: Harvard University Press.

Reuters (2019) 'Reuters Digital News Report'. Available at http://www.digitalnewsreport.org/ (accessed 20 August 2019).

Ribble, M. (2004) 'Digital Citizenship: Addressing Appropriate Technology Behavior', *Learning and Leading with Technology*, 32(1), pp. 6–11.

18 Re-imagining the democratic public

Stephen Coleman

We live in an era in which angry political discourse has become the norm. Major nations are increasingly run by men such as Erdoğan, Orbán, Modi, Putin, Duterte and Trump, who speak in capital letters and thrive on orchestrated outrage. The President of the United States is so angry that he has to be connected throughout his waking hours to an anger machine (Twitter) into which leak dregs of bitterness that vitiate the public sphere. Indignation is not confined to the political elite. A 2018 Pew study (Pew Research Center, 2018) asked social media users how the content they encountered tended to leave them feeling: 71 per cent chose the word 'angry'. A 2018 Pew analysis of congressional Facebook pages (Pew Research Center, 2018) found that the 'anger' emoticon is now the most common public response to posts by members of Congress. The UK Committee on Standards in Public Life (2017: 27) has reported that 'it is clear that intimidatory behaviour has become a significant and damaging feature of public life', and that:

> social media has sparked a step-change in the abuse and intimidation MPs, candidates, and others in public life receive. The instantaneous and direct nature of communication online has shaped a culture in which the intimidation of candidates and others in public life has become widespread, immediate, and toxic. This is exacerbated by the ability to hide behind the anonymity of social media profiles. (ibid.: 46)

Politicians have long been aware that arousing negative emotions is a forceful mobilizer of loyal support (Brader, 2005; Kosmidis et al., 2019). Angry voters are not merely primed to support their leader, but to bring down rival candidates for power and seek retribution against 'undeserving' or 'unbelonging' fellow citizens. The short-term effects of mass political anger have a catalysing potential; often, as we have seen in recent years, generating creative destruction by tearing down embedded structures of order and ditching institutional protocols. The longer-term effects are much more insidious, demobilizing collective action and diminishing citizens' sense of political efficacy. Mass public anger has a negative effect on information-seeking and desire to learn

(MacKuen et al., 2010), makes people vulnerable to being taken in by high-risk strategies, justified in terms of negative emotion (Huddy et al., 2007), and engenders pernicious hostilities that erode social solidarity. In short, the long-term consequences of creating a digital environment in which anger is the dominant tone are deeply corrosive to democratic norms.

Seeking to change the enraged tone of contemporary democracy is more than a merely ethical or aesthetic ambition. It should not be understood as a pious call to citizens to communicate more politely or to the corporate rulers of digital platforms to re-invent themselves as promoters of deliberative speech norms. Rather, it entails a willingness to re-imagine radically what constitutes political form. By 'form' I mean the cluster of core features that configure patterns of practice and experience into determinate social entities. For example, governing is a social practice, but government assumes a specific form, characterized by structural norms which can be cultivated in their own right, separately from the contingencies of practice. Practices of governing are expected to adhere to the formal norms of government. Rethinking how we do politics involves confronting tensions between the forms of being political that have assumed ontological naturalness over time and practices of acting politically that emerge and evolve through creative agency. I think that this is what John Dewey (1927 [1954]) had in mind when he wrote that to 'form itself the public has to break existing political forms'. Social scientists – and especially political scholars – tend to speak of 'the public' as a formed entity, an aggregate body that can be relied upon to play certain roles. (The underspecified role of the public in the famous politician–journalist–citizen pyramid upon which most political communication research is founded is a good example of how formalism constrains agency.) Dewey's argument is that for the public to exercise agency it must resist formal functions ascribed to it by others and create its own practices of acting politically. In the context of social media publics, this would entail refusing to be cast as a knee-jerk angry mob. Rather than accepting the form of digital political space on the terms that it is offered to them, they would break such forms and refunction (to use a key Brechtian term) their democratic environment.

Breaking with political formalism entails rejecting thinly institutionalist and procedural characterizations of democracy. It is not because politicians sit in parliaments or congresses, or elections between rival candidates take place, or citizens are free to curse the government, that democracy can be said to exist. Democracy might be secured through constitutional mechanisms, but it is not formed by them. The foundational gesture of democracy is the intersubjective coming together of citizens within whatever forms of mutual presence they are able to generate. As Hannah Arendt (1958: 198) famously stated, democratic publics emerge when people appear before one another 'where I appear to others as others appear to me'. She goes on to state that:

> The space of appearance comes into being wherever men are together in the manner of speech and action, and therefore predates and precedes all formal constitution of the public realm and the various forms of government, that is, the various forms in which the public realm can be organized. (Arendt, 1958: 218–19)

The question that preoccupies Arendt is not about how citizens can realize their role as members of a preconstituted public – as social media users, mass-media audiences, partisan activists or dutiful citizens, for instance – but how people can find creative ways of making their appearance to one another, thereby 'becoming a public'. The difference between these two objectives is so stark as to suggest that the glib use of the term 'democratic public' in much of the research literature about digital politics is destined to cause ambiguity and confusion. From a normative perspective, there can be few more important contemporary challenges than to explore the various forms in which the public realm – and, by implication, the public itself – can organize and be organized.

Breaking things: the digital disruption of existing forms

At first glance, the hacking and modification of video games by their users might seem to have little relevance to the making of democratic publics. But there is one sense in which the work of 'modders' (as those who modify gaming software and hardware with a view to performing a function not originally conceived or intended by the corporate owners and designers) can be seen as trailblazers of digital political disruption. As Postigo (2008) argues, when modders 'change or "poach" original content', what they produce can be 'radically transformative, essentially abandoning the original narrative, the original characters and other initial content'. In this sense, 'Modding is a form of meta-gaming – playing games that play with the game systems' (Scacchi, 2010).

Much contemporary democratic activism could be well described as 'meta-activism': less an attempt to engage with the processes and mechanisms of the political order than to highlight their unequal and unjust foundations. For example, in her analysis of the global hacktivist group Anonymous, Gabriella Coleman (2018: 20) states: 'While Anonymous has not proposed a programmatic plan to topple institutions or change unjust laws, it has made evading laws and institutions seem desirable. It has enabled action at a time when many feel that existing channels for change are either beyond their reach or too corrupt'.

By adopting what Gabriella Coleman describes as 'weapons of the geek' such activists are enabled, however playfully or peripherally, to challenge the rationalities and test the boundaries of elite power. Instead of participating from within, they are acting from without. They are exposing the limits of public institutions by exploring what happens when the people whose interests they claim to be serving get access to them. WikiLeaks was a notable example of this happening, with globally momentous consequences.

Another digital project that can be seen as having disrupted the official provision of political information is TheyWorkForYou.com, a website that makes available information about elected representatives' voting records, expenses and speeches in the United Kingdom Parliament as well as the Scottish Parliament and Northern Irish Assembly. The 200,000–300,000 people who access the site each month are able to annotate written parliamentary proceedings and create customized newsfeeds about the latest appearances of individual members, as well as receiving email alerts on any item mentioning certain keywords. They also have access to video recordings of debates in the House of Commons which can be searched using verbatim, time-stamped transcripts. This is a remarkably successful democratic tool, allowing citizens to take time to explore the ways in which they are being represented – or misrepresented. By creating a new form of critical scrutiny that uses metadata to interrogate representative claims (Saward, 2010), it can be said to be playing games with the political system.

Projects such as Anonymous and TheyWorkForYou invite people to exploit digital affordances with a view to disrupting the tight grip of elites upon existing forms of political information and communication. These projects at the very least place some strain on hegemonic forms, and at best focus minds upon the obsolescence of centralized models of political power. In being characterized as forms of digital disruption, they are seen as creating 'a type of environmental turbulence' resulting in 'the erosion of boundaries and approaches that previously served as foundations for organizing' particular social activities (Skog et al., 2018). In short, like the mood of anger that pervades contemporary political discourse, these projects run against the normative grain; they stir things up, but do not necessarily settle anything. The question of how these disruptive, tempestuous tendencies might generate enduring practices of democratic agency remains to be answered.

In a refreshingly critical analysis of 'political communication in a time of disrupted public spheres', two of the leading scholars in the field, Lance Bennett and Barbara Pfetsch (2018: 243), have observed that 'democracies today are experiencing various forms of legitimacy crises'. They offer an astute critique of the conceptual legacy of functionalism within political communication studies and suggest that 'bringing politics and democracy back into the forefront of the field are crucial to better understanding communication in fragmented public spheres, weak legacy media systems, and disrupted democracies' (ibid.: 250). It is one thing to improvise digitally disruptive irritants and cultivate tones of exasperated anger intended to shake up the 'old guard', but quite another to enroot habits of civic communication that vitalize democracy. It goes without saying (or should) that there is no techno-deterministic correlation between the digitalization of political communication and enhanced public agency. But neither should we accept the defeatist position that socio-technical affordances are mere snares that can only ever consolidate hierarchical power (Morozow, 2011). Political change begins with imaginative vision (Fischer and Gottweis, 2012; Wolf and Van Dooren, 2017), and the first step towards restoring the demos to an efficacious and dignified role within democracy is to think imaginatively about the available resources of regeneration.

If Bennett and Pfetsch's worthy objective of bringing democracy into the forefront of the field is to be realized, research will need to focus not only upon the rational-cognitive dimensions of political expression, but also upon the affective tonalities that spur or retard democratic form. Specifically, in the context of our current impasse, the question of how the energy that is currently being invested in the perpetuation of mass anger can be redirected to constructive ends must surely be one of the most urgent questions facing political communication scholars.

From anger to agency

This 'age of anger' (Mishra, 2017), in which political discourse is so commonly expressed through splutters of irritability within a lexicon of rancour, might also be characterized as an era of arrested agency. For, rather than the exercise of anything resembling democratic agency, the rage being vented appears to be a frustrated response to social forces that people feel are beyond their control.

Anger has its own affective history, which has veered over the centuries between encouragement to purgative emotional ventilation, as in the case of medieval peasants who 'might get enraged, but under normal circumstances

. . . were helpless to do much about it' (Freedman, 1998: 172), and puritan appeals to curb passion with a view to conforming to the prescribed 'feeling rules' of civility (Hochschild, 1979; Goffman, 1983; Stearns, 1994). While Peter Lyman (2004: 134) is right to point out that anger is often a response to injustice and should therefore not be dismissed 'as a psychological problem rather than as a form of political speech', expressions of political anger tend to be better at pinpointing unfairness than proposing what can be done about it. In short, angry talk is typically an affective surrogate for agency. For all of their rhetoric about returning control to 'the left behind', populists' splenetic narratives appeal to a plaintive trope of lost or stolen agency.

The role of digital media in the circulation of affects such as anger is too often rather simplistically explained. The fashionable view that social media are spaces of affective contagion emanates from discredited media-effects theory. Imagining that terrorists, racists, child abusers, misogynists and misanthropes have found through the Internet a new freedom to set the tone of political discourse misses the point that messages are impotent unless they are positively received. Trump's crazy tweets do not create the self-righteous fury of angry white bigots; they appeal to it. For, as Zizi Papacharissi (2015: 2) suggests in her astute study of affective circulation, hashtags 'serve as empty signifiers that invite ideological identification of a polysemic orientation'. It is not that a political feeling is formed and then disseminated, but that through the process of circulation, feelings come to be formed. Indeed, to form a binding collective feeling is to assemble a public (see Warner, 2002). Papacharissi (2015: 9) refers to these as 'affective publics', characterized by shared 'feelings of belonging and solidarity, however fleeting or permanent those feelings might be'. Being indignant is the most common contemporary form – the generic performance, one might say – that people adopt when they encounter injustice. The content and tone of popular anger serves to call attention to an absence of agency: 'We're bound to lose – be ignored – be betrayed – be sneered at.' The question for research is how such desolate refrains in response to injustice might be transformed into a vocabulary through which justice can be spoken of as a feasible aspiration.

Pursuing such research entails understanding more about the mediation of shared political narrative. In a neoliberal era that prides itself on justifying decisions by nullifying ordinary experiences and voices, being able to tell one's own story and share it with others constitutes an immensely valuable form of agency. However, within a highly competitive attention economy, it is one thing to be able to tell a story and another to have it heard by enough other people to make a difference. Building on the insightful work of Bennett et al. (2018) on 'negotiating attention and meaning in complex media ecologies', there is important research to be conducted on how ideas and feelings travel within and across social networks and come to be mediated by frames of meaning. This involves asking critical questions about how digital platforms not only structure and capture interaction, but also orient users to particular understandings of what it feels like to be within those interactive relationships. As Stefania Milan (2018: 514) puts it in her excellent study of digital traces, 'Social actors are "transmuted" into an ever-changing multitude of posts, pictures, videos, and emotions expressed in varied formats among those enabled by a given platform – including via dedicated buttons and emoticons.'

The terms upon which people encounter one another online determines their readiness (or unfitness) for democratic action. Sometimes, as when Facebook irresponsibly experimented with trying to control the emotions of nearly 700,000 of its users (Kramer et al., 2014; Panger, 2016), affective manipulation is intentional, but generally it is more a case of what Margaret Wetherell (2015: 160) calls 'entanglement': 'affective practice is a moment of recruitment, articulation or enlistment when many complicated flows across bodies, subjectivities, relations, histories and contexts entangle and intertwine together to form just this affective moment, episode or atmosphere with its particular possible classifications'. Viral moods of casual anger serve to define global and local situations that are bewildering, frustrating and deeply affecting through a common cry of pain. Much of what passes for enraged digital political discourse is not reflective opinion, but cathartic wailing. It is what people do when they think they cannot do anything. It is a vociferous declaration of withered agency.

The current crisis of agency raises a formidable challenge for researchers who care about the health of the political democracy they study. It entails rethinking political practices that have become ossified within obsolete forms dominated by slick marketing and cynical manipulation. It calls for normative abandonment of the lazy assumption that the best we can hope for from politics online is an unchecked anger-fest. The challenge is to re-imagine spaces of political communication, as if humans were capable of more than shouting at one another.

For this to happen, democracy requires a new discourse architecture (Freelon, 2015) that maximizes extensive social connectivity (Bennett and Segerberg, 2013), exposes online information-seekers to unsought news and knowledge (Valeriani and Vaccari, 2016; Weeks et al., 2017; Matthes et al., 2019), encourages citizens to transcend homophilic silos by engaging in cross-cutting discussion (Mutz, 2002; Bakshy et al., 2015), ensures that elected representatives are driven by the informed and reflective demands of citizens (Coleman and

Blumler, 2009; Coleman, 2017), and values working through problems rather than falling back upon knee-jerk ideology (Parkinson and Mansbridge, 2012; Stromer-Galley et al., 2015; Coleman, 2018). Research and practice of this kind contributes to an understanding of what Dewey referred to as the public forming itself by breaking existing forms.

But beyond structures of political connection, which is where digital technologies have tended to be seen as playing a reconfiguring role, there are the capabilities that people bring to democracy. People who do not feel confident in expressing their political views and values in everyday offline encounters are the least likely to become political agenda-setters online (Kim, 2015; Halpern et al., 2017). It is therefore important for researchers to explore links between the capabilities that people say they need in order to become the kind of democratic citizens they would like to be, and the available opportunities and affordances for realizing such capabilities (Moss, 2018; Coleman et al., 2018). In politics, as elsewhere in life, people who feel frustrated by lack of capabilities they think they should possess are likely to become angry. A pressing question for research concerns how people move from the ire of disappointed expectations (of themselves as well as the political system) to undaunted agency.

References

Arendt, H. (1958) *The Human Condition*. Chicago, IL, USA and London, UK: University of Chicago Press.

Bakshy, E., Messing, S. and Adamic, L.A. (2015) 'Exposure to Ideologically Diverse News and Opinion on Facebook', *Science*, 348(6239), pp. 1130–32.

Bennett, W.L. and Pfetsch, B. (2018) 'Rethinking Political Communication in a Time of Disrupted Public Spheres', *Journal of Communication*, 68(2), pp. 243–53.

Bennett, W.L. and Segerberg, A. (2013) *The Logic of Connective Action: Digital Media and the Personalization of Contentious Politics*. New York: Cambridge University Press.

Bennett, W.L., Segerberg, A. and Yang, Y. (2018) 'The Strength of Peripheral Networks: Negotiating Attention and Meaning in Complex Media Ecologies', *Journal of Communication*, 68(4), pp. 659–84.

Brader, T. (2005) 'Striking a Responsive Chord: How Political Ads Motivate and Persuade Voters by Appealing to Emotions', *American Journal of Political Science*, 49(2), pp. 388–405.

Coleman, E.G. (2018) 'Logics and Legacy of Anonymous'. In J. Hunsinger, M.M. Allen and L. Klastrup (eds), *Second International Handbook of Internet Research*. Dordrecht: Springer, pp. 1–22.

Coleman, S. (2017) *Can the Internet Strengthen Democracy?* Cambridge: Polity Press.

Coleman, S. and Blumler, J.G. (2009) *The Internet and Democratic Citizenship: Theory, Practice and Policy*. Cambridge: Cambridge University Press.

Coleman, S., Moss, G. and Martinez-Perez, A. (2018) 'Studying Real-Time Audience Responses to Political Messages: A New Research Agenda', *International Journal of Communication*, 12, p. 19.

Committee on Standards in Public Life (2017) *Intimidation in Public Life: A Review by the Committee on Standards in Public Life*. Controller of Her Majesty's Stationery Office, UK.

Dewey, J. (1927 [1954]) *The Public and Its Problems*. Denver, CO: Swallow Press.

Fischer, F. and Gottweis, H. (eds) (2012) *The Argumentative Turn Revisited: Public Policy as Communicative Practice*. Durham, NC, USA and London, UK: Duke University Press.

Freedman, P. (1998) 'Peasant Anger in the Late Middle Ages'. In B.H. Rosenwein (ed.), *Anger's Past: The Social Uses of an Emotion in the Middle Ages*. Ithaca, NY: Cornell University Press, pp. 171–90.

Freelon, D. (2015) 'Discourse Architecture, Ideology, and Democratic Norms in Online Political Discussion', *New Media and Society*, 17(5), pp. 772–91.

Goffman, E. (1983) 'The Interaction Order: American Sociological Association, 1982 Presidential Address', *American Sociological Review*, 48(1), pp. 1–17.

Halpern, D., Valenzuela, S. and Katz, J.E. (2017) 'We Face, I Tweet: How Different Social Media Influence Political Participation through Collective and Internal Efficacy', *Journal of Computer-Mediated Communication*, 22(6), pp. 320–36.

Hochschild, A.R. (1979) 'Emotion Work, Feeling Rules, and Social Structure', *American Journal of Sociology*, 85(3), pp. 551–75.

Huddy, L., Feldman, S. and Weber, C. (2007) 'The Political Consequences of Perceived Threat and Felt Insecurity', *Annals of the American Academy of Political and Social Science*, 614(1), pp. 131–53.

Kim, B.J. (2015) 'Political Efficacy, Community Collective Efficacy, Trust and Extroversion in the Information Society: Differences between Online and Offline Civic/Political Activities', *Government Information Quarterly*, 32(1), pp. 43–51.

Kosmidis, S., Hobolt, S.B., Molloy, E. and Whitefield, S. (2019) 'Party Competition and Emotive Rhetoric', *Comparative Political Studies*, 52(6), pp. 811–37.

Kramer, A.D., Guillory, J.E. and Hancock, J.T. (2014) 'Experimental Evidence of Massive-Scale Emotional Contagion through Social Network', *Proceedings of the National Academy of Sciences*, 111(24), pp. 8788–90.

Lyman, P. (2004) 'The Domestication of Anger: The Use and Abuse of Anger in Politics', *European Journal of Social Theory*, 7(2), pp. 133–47.

MacKuen, M., Marcus, G., Neuman, W.R. and Miller, P.R. (2010) 'Affective Intelligence or Personality? State vs. Trait Influences on Citizens' Use of Political Information', Paper presented at the annual meeting of the American Political Science Association and the New York Columbia University Political Psychology Colloquium.

Matthes, J., Knoll, J., Valenzuela, S., Hopmann, D.N. and Von Sikorski, C. (2019) 'A Meta-Analysis of the Effects of Cross-Cutting Exposure on Political Participation', *Political Communication*, pp. 1–20. doi:10.1080/10584609.2019.1619638.

Milan, S. (2018) 'Political Agency, Digital Traces, and Bottom-Up Data Practices', *International Journal of Communication*, Special Section, 'Digital Traces in Context', edited by Andreas Hepp and Andreas Breiter, 12, pp. 507–25.

Mishra, P. (2017) *Age of Anger: A History of the Present*. London: Macmillan.

Morozow, E. (2011) *The Net Delusion: The Dark Side of Internet Freedom*. New York: Perseus Book Group.

Moss, G. (2018) 'Media, Capabilities, and Justification', *Media, Culture and Society*, 40(1), pp. 94–109.

Mutz, D.C. (2002) 'The Consequences of Cross-Cutting Networks for Political Participation', *American Journal of Political Science*, 46(4), pp. 838–55.

Panger, G. (2016) 'Reassessing the Facebook Experiment: Critical Thinking about the Validity of Big Data Research', *Information, Communication and Society*, 19(8), pp. 1108–126.

Papacharissi, Z. (2015) *Affective Publics: Sentiment, Technology, and Politics*. New York: Oxford University Press.

Parkinson, J. and Mansbridge, J. (eds) (2012) *Deliberative Systems: Deliberative Democracy at the Large Scale*. New York: Cambridge University Press.

Pew Research Center, Internet and Technology (2018) 'Public Attitudes Toward Computer Algorithms'. Available at https://www.pewinternet.org/2018/11/16/public-attitudes-toward-computer-algorithms/ (accessed 5 September 2019).

Postigo, H. (2008) 'Video Game Appropriation through Modifications: Attitudes Concerning Intellectual Property among Modders and Fans', *Convergence*, 14(1), pp. 59–74.

Saward, M. (2010) *The Representative Claim*. Oxford: Oxford University Press.

Scacchi, W. (2010) 'Computer Game Mods, Modders, Modding, and the Mod Scene', *First Monday*, 15(5). Available at http://firstmonday.org/article/view/2965/2526/ (accessed 5 September 2019).

Skog, D.A., Wimelius, H. and Sandberg, J. (2018) 'Digital Disruption', *Business and Information Systems Engineering*, 60(5), pp. 431-7.

Stearns, P.N. (1994) *American Cool: Constructing a Twentieth-Century Emotional Style*. New York: NYU Press.

Stromer-Galley, J., Bryant, L. and Bimber, B. (2015) 'Context and Medium Matter: Expressing Disagreements Online and Face-to-Face in Political Deliberations', *Journal of Public Deliberation*, 11(1), p. 1.

Valeriani, A. and Vaccari, C. (2016) 'Accidental Exposure to Politics on Social Media as Online Participation Equalizer in Germany, Italy, and the United Kingdom', *New Media and Society*, 18(9), pp. 1857–74.

Warner, M. (2002) *Publics and Counterpublics*. New York: Zone Books.

Weeks, B.E., Lane, D.S., Kim, D.H., Lee, S.S. and Kwak, N. (2017) 'Incidental Exposure, Selective Exposure, and Political Information Sharing: Integrating Online Exposure Patterns and Expression on Social Media', *Journal of Computer-Mediated Communication*, 22(6), pp. 363–79.

Wetherell, M. (2015) 'Trends in the Turn to Affect: A Social Psychological Critique', *Body and Society*, 21(2), pp. 139–66.

Wolf, E.E.A. and Van Dooren, W. (2017) 'How Policies Become Contested: A Spiral of Imagination and Evidence in a Large Infrastructure Project', *Policy Sciences*, 50(3), pp. 449–68.

Index